8

The Benjamin Franklin Lectures

OF THE

University of Pennsylvania

FOURTH SERIES

The Scientists Look at Our World

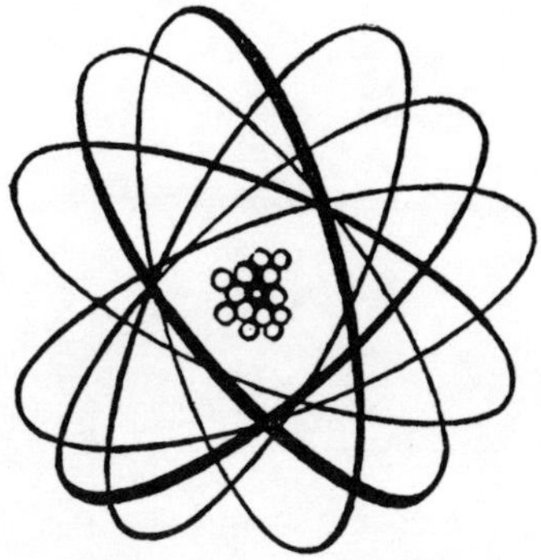

Essay Index Reprint Series

BOOKS FOR LIBRARIES PRESS
FREEPORT, NEW YORK

Library of Congress Cataloging in Publication Data
Main entry under title:

The Scientists look at our world.

(Essay index reprint series)
Original ed. issued as 4th ser., 1951, of The Benjamin Franklin lectures of the University of Pennsylvania.
1. Science--Addresses, essays, lectures. I. Houston, William Vermillion, 1900- II. Series: Pennsylvania. University. Benjamin Franklin lectures, 1951.
[Q171.S43 1972] 508'.1 71-142694
ISBN 0-8369-2859-8

PRINTED IN THE UNITED STATES OF AMERICA

The Scientists Look at Our World

W. V. HOUSTON

W. ALBERT NOYES, JR. • *CURT STERN*

ALAN GREGG • *WENDELL H. CAMP*

Introduction by JOHN M. FOGG, JR

Contents

The Scientists Look at Our World

Introduction

In inviting five distinguished scientists, from as many fields of specialization, to deliver the 1951 Series of Benjamin Franklin Lectures, the Lecture Committee of the University of Pennsylvania had several aims in mind. First, it requested each of these eminent authorities to discuss the advances achieved in his discipline in recent years and evaluate the position which it occupies at mid-century. Second, it was hoped that each speaker would touch upon the interdependence of the several fields of science and would emphasize the need for breadth as well as depth in scientific education today. Finally, it was suggested that our speakers address themselves to the proposition that now more than ever before it is necessary for the scientist to be aware of his responsibility to society.

How well our speakers have performed the tasks assigned to them may be judged by the reader of this volume. As the series of lectures developed, it became increasingly evident to our audiences that a hard central core of agreement was emerging and that although each speaker dealt, sometimes in a rather technical manner, with the subject matter of his own field, a number of basic principles, attitudes, and judgments characterized their views and were expressed by each in turn.

One had a right to expect from these five leaders a broad

and penetrating analysis of recent advances and present trends in their respective fields, and each of them amply fulfilled this role. That which might not have been predicted with equal certainty, however, was their common appreciation of the relationships which exist among the various fields of science, as well as the relationships between the scientist and the man in the street. To cite but a single example, the discovery of nuclear fission and the invention of its Frankenstein monster, the atom bomb, lie in that borderline where physics and chemistry meet and overlap. Yet each of our speakers was concerned with the social, political, and economic, as well as the scientific, implications of this discovery.

Each emphasized also that the techniques developed in one field are frequently applied to another. Thus we have the botanist calling attention to a need for an understanding of the role of tracer elements in plant nutrition or underlining the role which the physical chemist must play in throwing light on problems of colloidal balances in the soil. The medical man makes a plea for the application of genetic techniques to human problems and stresses the need for an ecological approach to the study of medicine. The student of genetics invokes the aid of the physicist and biochemist in the solution of his problems, and the chemist readily admits that the application of the techniques of the physicist has become a useful tool in his own laboratory.

All of our speakers called attention to man's unenviable record of wastefulness in utilizing his natural resources. Yet there is general agreement that man possesses the inherent capacity to profit from his mistakes, as well as the ingenuity and vision which, if properly applied, will enable him to deal effectively with his environment.

Finally, our speakers have made it crystal clear that the scientist, whether he be a teacher in a university or a member of a research institution, can no longer work entirely in the cloistered seclusion of a bygone era. He must descend from his ivory tower to acquaint himself with what his colleagues in other fields of science are doing and, above all, he must entertain and nourish a realization of the fact that society looks to him for leadership in the intelligent application of those forces which he has liberated.

JOHN M. FOGG, JR.

W. V. HOUSTON

Description of the Physical World

Until recently physics was almost the least known of the natural sciences. Certainly it was relatively unknown to the great mass of people. About fifteen years ago I was introduced to an apparently friendly and expectant audience, as a "Professor of Psychics." The audience was evidently better prepared for this eventuality than I was, and the chairman seemed to think it an entirely appropriate interpretation of the apparatus I had brought along for demonstration.

But within the last ten years physics has moved from a position of relative obscurity to one of uneasy prominence. Almost every literate person has heard the word and has some kind of idea of what a physicist is. Many of the ideas, of course, border on fantasy. The feeling that the physicist goes around with an atomic bomb in his hip pocket may have induced a curious kind of respect for the profession, but has not given it a reputation as a beneficent element of society.

Many physicists believe that a wider dissemination of knowledge concerning the facts of physics, the attitudes and motivations of physicists, and the real means by which advances in physics are made, would not only put the well-meaning scientist in a slightly less harsh light, but would at the same time contribute to the general welfare.

The outward manifestations of physics are material things. When speaking of physics, the man in the street naturally thinks of radios, of television, of airplanes, of electric lights, and above all in these days, of atomic or nuclear bombs.

We each get a trifle weary of hearing that this is a scientific age. But the assertion means far different things to different persons. Some people refer to the material equipment in and with which we live, to the air-conditioning machines, to the multifarious gadgets that assist us on every hand. And these are indeed marvels. But science is not, obviously, the gadget. One has to look behind the manifestation, to the concept.

Probably most of you have ridden today in a vehicle beyond the wildest hopes of Benjamin Franklin. On a luxuriously upholstered seat you were protected from the wind and possible rain, and you were wafted with scarcely a jar over distances that not long ago represented a major expedition. And how did this come about? Few of you were concerned that less than thirty inches in front of you was an inferno of exploding gasoline, that the heat was so tremendous as to require a constant stream of water to keep the steel from melting. Few of you were alarmed that a few inches behind you was a tank of fuel, sufficient to propel you 200 miles ahead if properly applied, or 200 feet into the air if improperly applied. The marvel of the automobile is not so much that it goes, but that it goes quietly, surely, dependably, in hot weather or cold, in rain, snow, and wind. It is a marvel of reliability and usefulness.

But although these material things are manifestations

of physics, and the corresponding engineering techniques are essential in the further development of physics, it is clearer each day that the central core of physics is a group of ideas. It is a group of ideas concerning the nature of the physical world. The development, the interpretation, and the application of these ideas constitute the progress of physics.

There seem to be two basic ways of thinking about the physical world. These were formulated in ancient times, and in fundamental concept have not changed much since. The first is to picture the world as what is called a "continuum." This is simply an idealization of our experience with solid bodies. We have contact with distant objects because the space between is filled with some kind of substance. We see light because the disturbance travels through an intervening medium. In his excellent book on the history of light and electricity during the eighteenth and nineteenth centuries, E. T. Whittaker pointed out the strong influence of the philosophy of Descartes in support of the idea of an all-pervading ether. During the nineteenth century the ether was endowed with some very remarkable properties. It was, for instance, thought to have infinite rigidity; yet meanwhile it was supposed to offer no resistance to the motion of bodies through it. We still look to the ether as the seat of electrical phenomena, as the carrier of radio waves and of light; and we still picture it as having a continuous structure and nothing of a granular nature. The more sophisticated have now abandoned the word "ether" and refer to it as the "electromagnetic field," but they do about the same things with it as the nineteenth-century philosophers did with their ether.

The second method of looking at the world was also formulated by the Greeks. Democritus saw all matter as composed of atoms or individual particles moving around with high velocity, and conveying impressions from one spot to another by their own actual movement. This point of view lay dormant for a long time, was revived in the seventeenth century, and during the nineteenth century became the basis of the study of chemical reactions. The extreme exponents of the atomistic philosophy would sometimes suggest even an ether as a substance composed of very small atoms. So imbued were they with the power and universality of this formula, that although no one had ever seen an atom, there was no hesitancy in applying the same conception to immaterial ideas such as the soul. Nevertheless, the evidence for atoms was largely circumstantial; and a strong group of distinguished men maintained in the last years of the nineteenth century that although material atoms might be a useful figment of the imagination, they must not be given too realistic an interpretation.

At the end of the nineteenth century, then, the ordinary physicist who was not too much concerned with philosophic consistency saw his science divided into two parts: on the one hand, the physics of the ether described the propagation of light and the electrical and magnetic phenomena with which he was acquainted; on the other hand, the physics of matter, vaguely based on an atomic philosophy, although largely expressed in macroscopic terms, described the phenomena of solid, liquid, and gaseous bodies.

Against such a background, the first half of the twentieth century has seen tremendous development, not only in the application of physics, but in its ideas. There seem today to be three principal groups of ideas that form the core of contemporary physics, and I should like to discuss them with you briefly. When Einstein, in 1905, announced his special theory of relativity, he put the capstone on forty years of work by various mathematical physicists. The consequences of his insight are a major factor in the thinking of physicists today. In the second place, the twentieth century has seen the development of the atomic particle idea. It has seen means develop for making evident the effects of individual atoms, has seen matter subdivided until at one time everything seemed to be composed of electrons and protons. It has seen later further analysis develop neutrons, positive electrons, neutrinos, mesotrons of various kinds, and open the field to possibly an unlimited number of different things now described as particles. And in the third place, the quantum mechanics, introduced by Planck in the last month of the year 1900, has developed into a philosophy pervading the whole of physics. It makes clear that although we refer to atoms as particles and to the electromagnetic field as a continuum, simple generalizations of our sense perceptions are inadequate to describe the behavior of these things. Much more subtle means must be used to interpret the experimental facts on which physics is based.

Einstein's theory of relativity, of course, grew out of electrodynamics and the electromagnetic theory of light. When it became well established that light was a wave motion in the ether, it was natural to wonder about the motion of the earth through the ether and to undertake to measure it. Michelson and Morley devised their famous experiment, in order to measure the rate at which light was propagated, parallel to the direction of motion of the earth, and perpendicular to it. The astounding result was that the velocity is the same, no matter whether the laboratory is moving in the direction of the light or not. This seemed an incredible result. It was an incredible result. For years many people refused to believe it. It means, for example, that if this room could be moving away from the sun with the velocity of light, the light from the sun would, nevertheless, travel through this room at just the same speed as though the room were standing still.

It was the genius of Einstein to recognize the importance of this experimental result, and to accept it as an observational fact, without undertaking to understand it, or to explain it, in terms of earlier experience. He merely investigated the conclusions that must follow from it. One of these conclusions was the dependence of mass on velocity. He showed, for example, that as a particle is set into motion, its mass must increase, and that the energy of a particle may be thought of as consisting of a rest energy and an energy due to its motion. Early experiments on the mass of electrons tended to confirm this result, even though they were not highly precise. A more precise confirmation was given when Sommerfeld in 1916 applied this theory to the motion of an electron, around a proton, in a hydrogen atom. He showed that the observed hydrogen spectrum could be accounted for, if the mass of the electron was supposed to vary, in the way indicated by Einstein.

And then in 1916 Einstein extended his considerations to include gravitation, and developed the general theory of relativity. The general theory of relativity has been the stimulus for much mathematical ingenuity. Methods of tensor calculus have been developed and applied. It has been easy to conceal the ideas of the theory in this mountain of mathematical manipulation. Nevertheless, out of this work have come a few results confirmed by astronomical observations, such as the motion of the planet Mercury, and the bending of light around the sun. Further extensions of the theory, combined with hypotheses as to the nature of the universe, have given rise to problems which are now occupying the attention of astronomers everywhere. One of the purposes for which the 200-inch Hale telescope on Mt. Palomar was constructed was to investigate one of these problems, the density of nebulae out to the greatest possible distances.

The principal value of the general theory of relativity has been in suggesting astronomical problems and problems of cosmology. As far as I know, it has been of relatively little use in problems confined to the laboratory. Astronomers will always speculate and wonder as to the structure of the universe, whether it is bounded or infinite, whether it is all like the part in our vicinity, and how it originated. The general theory of relativity has given a new framework in terms of which these speculations can be formulated. The speculations themselves are as old as astronomy.

But probably the most significant consequence of the Einstein theory of relativity is the equivalence of mass and energy. Among the great accomplishments of the nineteenth century was the establishment of the law of conservation of energy and the law of conservation of mass. Conservation of mass was established only after the techniques of chemistry had becom quantitative. It served to focus attention on the mass of a substance as its fundamental property, and led to a new and more respectful attitude toward matter itself. Matter became clearly distinguished from ether, and from non-material things, and was easily raised to a central position in a materialistic philosophy.

The conservation of energy required still more refinement of experimental techniques to establish it. In fact, it was not a tenet of physics during Franklin's lifetime, and became generally accepted only about a hundred years ago. The idea of mechanical energy was, of course, inherent in Newton's work, and was elaborated by those who followed him. It took a hundred and fifty years more to extend the idea to heat energy in a quantitative way. Even now one occasionally meets an inventor of a perpetual motion machine who doesn't seem to believe it. But the law made such an impression on physicists that it was soon extended to include not only heat but electromagnetic phenomena, by an appropriate definition of electromagnetic field energy.

These two laws were two of the pillars on which physics rested at the end of the nineteenth century, but now we find there are not two pillars. There is only one. Mass and energy are not really different. They are different aspects of the same entity. One can be changed into the other, and only the sum is constant.

This equivalence has been accepted by physicists for a good many years, but its really striking demonstration came about through the release of nuclear energy. When an atom of uranium divides into two parts, the two parts together are not so heavy as the one from which they originally came. Mass is not conserved. The vanishing mass has been transformed into energy which is available to be used in a variety of ways. The development of atomic bombs and of other methods of using nuclear energy is the ultimate application of this law, which comes from Einstein's theory of relativity.

The work on relativity has had a great appeal for philosophers. It has put a very definite limitation to the Cartesian idea of an all-pervading ether. Even though one still sees light as a wave motion moving through something or other, this something or other is less physical than it was in the minds of the nineteenth-century philosophers. It is not something with reference to which motion can be detected. Its solidity as a frame of reference has evaporated.

Although the theory of relativity grew out of work in electrodynamics and provided solutions for certain of the problems, many problems still remain. Recent attempts to combine quantum mechanics and electrodynamics have produced again a tremendous amount of mathematics, a certain amount of physics, but relatively little feeling of a final result.

It seems not inappropriate to regard the present special theory of relativity, as it applies to electromagnetic phenomena, as merely a first approximation. One can formulate the theory of relativity in terms of the requirement that Maxwell's electromagnetic field equations must hold with reference to any observer. But Maxwell's equations are linear equations. In other fields of physics, we have cases where linear equations are merely first approximations as, for example, in the theory of sound. Sound travels with a well-defined velocity, dependent upon the elasticity and the density of the medium. No disturbance will travel faster than this velocity of sound, unless it becomes a very large disturbance, so large that the approximate equations no longer apply. It is then possible to have a shock wave, or highly energetic pulse of sound, traveling at a higher than the normal velocity.

In the electrodynamics of Maxwell and in the relativity of Einstein, the velocity of light is a limiting velocity. No energy can be transmitted at a higher speed, but may it not be true that Maxwell's equations are obeyed only by relatively weak electric and magnetic fields? May it not be that the highly intense fields in the neighborhood of electrons and nuclear particles obey different laws and are not subject to restrictions of Einstein's theory?

The second leading idea of the physics of the twentieth century is that of the atomic nature of matter. Something like 150 years ago the existence of chemical atoms was pretty well recognized. They were necessary to explain the laws of chemical combination and were regarded as identifiable and indivisible. During the nineteenth century the periodic table of the elements was invented, and it appeared that the discovery and identification of the elements might some time be complete. It appeared that the world was built of about ninety-three different kinds of atoms, and one foresaw the time when the properties of ordinary matter could be described in terms of these atoms and their interactions.

But one of the first things in the twentieth century was the production of negative electrons, and the discovery that they were apparently constituents of all kinds of matter, whatever its chemical constitution. It was with this discovery that the physicist entered into the problem of the atomic nature of matter. He broke up the chemical atoms and took out electrons. When Professor Millikan in 1910 showed conclusively that electric charge exists only in discrete units which he called electrons, the physicist had begun on his program of identifying fundamental particles.

In the original sense of the word, the electron appeared to be a true atom. It apparently was indivisible. All electrons were the same. They all had the same electric charge. They all had the same mass; and, as it developed later, they all had the same angular momentum and magnetic moment. It was quite convenient to visualize an electron as a small sphere, with a color depending upon the taste of the person imagining it, with a size which never could be quite tied down, and with a distribution of electricity, which it was never possible to firmly establish. But as time went on, it became clearer that possibly an electron really was a fundamental atom, that it therefore made no sense to talk about its shape, its size, or its constitution. An electron had an electric charge. It had a mass. It had an angular momentum. It had a magnetic moment. It had position and velocity. And these things seemed to justify the designation of it by the term "particle." Nevertheless, because it was a fundamental particle, its visualization had to be decidedly limited. The uninhibited physicist, elated by his success in analyzing chemical atoms, has always wanted to ask about the structure of an electron, but up to the present he has found no very satisfactory answer.

In 1912 Lord Rutherford showed that an atom consists of a heavy nucleus surrounded by numerous electrons. And so for a time one could think that the atom of the chemist had been fully analyzed into smaller parts, and that each atom consisted of a characteristic nucleus surrounded by a suitable number of electrons. This left a fairly satisfying picture. Instead of ninety-three elementary particles, one now had ninety-four—ninety-three different kinds of nuclei and one kind of negative electron. One could imagine all the phenomena of the material world to be describable in terms of these elementary particles.

But even in the early part of the century there was a small fly in the ointment. The discovery of radioactivity demonstrated that these atomic nuclei did not stay put. Occasionally one of them seemed to change to another. In some cases it split and gave out a helium nucleus, or alpha particle. In other cases it ejected an electron. But whichever process was used, a nucleus which originally belonged to one chemical element changed itself and became the nucleus of another chemical atom.

If a nucleus was not immutable, it presumably had a structure and a constitution. And so one undertook in his imagination to build up all nuclei from hydrogen nuclei, called protons, and electrons. By putting in enough protons, one could get the necessary mass; and by putting in electrons, get the necessary charge, without affecting the mass very much. Any resulting mass discrepancies were associated with an energy of binding according to Einstein's law of the equivalence of mass and energy.

And then, in the early 1930's, there was identified a particle having approximately the mass of a proton, but having no charge and so designated as a neutron. And so there soon developed the idea that a nucleus was made up of protons and neutrons. We now have the picture of atomic nuclei made up of protons and neutrons, and we undertake to understand what holds them together, why some are stable while others decay, why some have different angular momenta from others, why some have magnetic moments and electric moments. And one is hoping that the near future may provide laws of force between these nuclear particles, which will enable the structure of the individual nuclei to be quantitatively determined.

Studies of this kind have led to a kind of nuclear chemistry. Different kinds of nuclei can be combined to give still others. Some reactions take place easily, some with difficulty. Some give off energy, while energy must be added to others to make them go. A general survey of the nuclei shows that those of the lighter elements will combine with a release of energy. The remaining mass is less than the mass originally present in the constituent particles. The heavy elements will not combine in this way, for the mass of the combined nucleus will be greater than of the constituents; and this extra mass must be provided in the form of energy. A corollary, however, is that heavy elements should tend to break up into lighter ones, and many of them do. Radium, polonium, and others decay radioactively, and many others break up when appropriately stimulated. As everyone knows nowadays, uranium 235 breaks up into two approximately equal fragments when it is set off by collision with a slow neutron. This reaction has become famous, not only because of the amount of energy released, but because the reaction itself produces enough neutrons to stimulate fission in other atoms, and the process goes on into a chain reaction.

But at the other end of the periodic table, nuclear reactions are possible also. If three helium atoms could be shoved together the result would be carbon, with some of the mass going off in the form of energy. If two protons and two neutrons could be brought close enough together the result would be helium and an explosive amount of energy. Just as hydrogen and oxygen, mixed in the right proportions and heated hot enough at one point, explode to give water, so hydrogen and neutrons may be ignited to give an explosion and helium.

Stated in this way it sounds easy, and one might wonder why all the study and research and, in particular, why all the money the government is reported to be pouring into the problem of creating a hydrogen bomb? But in spite of the simplicity of statement, the technical problems associated with practical applications of nuclear energy are numerous and difficult. Nevertheless, they can be, and are being, solved.

As soon as techniques were developed for breaking up atomic nuclei, all sorts of other things began to appear. One of the first of these was designated as a neutrino. Sometimes a radioactive nucleus emits an electron, a so-called beta ray. A careful measurement of the energy of the nucleus before and after the emission of a beta ray, together with the measurement of the energy of the emitted electron, fails to account for all the energy. Some apparently has disappeared and must be found. The conservation of energy and mass is so firmly established as a part of physics that when an apparent violation of the law appears, every effort is made to rescue it, before giving a thought to the possibility that the law may be wrong. Hence, one postulates the simultaneous emission of a particle called a neutrino along with the electron. This neutrino has curious properties. When at rest, it has a zero mass, or at least a very nearly zero mass. This is merely another way of saying that when at rest, it really doesn't exist. All the mass it has is due to energy given to it when it is ejected.

Numerous physicists have put in a good deal of time and energy trying to detect the momentum which a neutrino carries off, but the results have been rather inconclusive. Apparently this phantom particle is called a particle more or less by courtesy, and is really the ghost of the conservation of energy.

Again in the late 1920's Carl Anderson observed in cosmic rays a particle with the mass of an electron but with a positive charge. This positive electron, or positron, appears to have only a transient existence. It tends to combine with any nearby negative electron and disappear with the production of two gamma rays. Positive electrons also appear as products of nuclear reactions when sufficient energy is available. But they must be studied quickly. They cannot be kept for leisurely observation.

Further studies of cosmic rays have led to the identification of particles intermediate in mass between electrons and protons, and designated as mesotrons or mesons. For a number of years efforts were directed toward measuring the mass of a mesotron, and it has finally been concluded that there are at least two kinds of them, and, in fact, there may be more. And so instead of protons and negative electrons, we again have begun to build up a list of so-called fundamental particles, or atomic constituents, of a variety of kinds; and we are far from understanding their interactions and the relationships between them.

The original idea of the ancient Greeks was to find in a number of immutable atoms some permanence in what appeared to be a rapidly changing world. At least the atoms themselves were not supposed to change, they merely changed their relationships to each other. Our new atoms, however, do not satisfy these conditions. They are not stable. They tend to disappear. They change one into another. They split up and combine. They are not very satisfactory atoms, and yet they provide the only reasonable picture which we now have of the ultimate constitution and structure of matter.

The third major group of ideas of the physics of the twentieth century constitutes quantum mechanics. It is really a twentieth-century product. The first introduction of a quantum was given in a paper by Planck in December 1900; and the fiftieth anniversary of this occasion was celebrated at the meeting of the American Association for the Advancement of Science in Cleveland last December.

Quantum mechanics took twenty-five years to develop, and I shall not undertake to outline the painful and laborious steps by which it attained its present formulation. Instead, I shall mention only a few of its characteristic features and some of the implications that seem to follow from it.

Quantum mechanics is an interesting and curious mixture of the ideas of particles and of a continuum. One of its underlying experimental facts is the phenomenon of the Compton effect. In the interaction between light and electrons, or any other particles, it is found that the light itself must be treated as a particle. It must be ascribed a certain energy and momentum and the property of interacting with another particle all at once. There are plenty of experiments on diffraction and interference to make it clear that light and X-rays are waves, and it is easy to think of them as waves in the ether or in the electromagnetic field. But, on the other hand, it is just as clear from the

Compton effect that light also has the properties of particles.

In the second place, it has been shown experimentally that particles such as electrons and protons, and even whole nuclei, are scattered by a diffraction grating, just as is a train of waves. The experiment must be described and analyzed in terms of a wave length. It is also just as certain as can be that an electron has a definite charge and a definite mass, and one likes to think of it as being in a definite place. Nevertheless, if a beam of electrons falls on a diffraction grating, the diffraction pattern is just as informative as that produced by a beam of X-rays or a beam of light.

Apparently both light and electrons are sometimes particles and sometimes waves. The distinction is not between different entities, but between different ways of looking at the same thing.

The names of Heisenberg, Schroedinger, and Dirac are those principally associated with the mathematical formulation of the theory that will include all of these facts. I shall not undertake to discuss the nature of the formulation here; but out of it and the attempt to interpret it have come a number of semi-philosophical principles and ideas. One of these is known as Heisenberg's principle of indetermination. Heisenberg pointed out that as a matter of experimental fact, if one tries to measure the position of an electron by any known means, the act of measurement will so interfere with the electron's motion as to invalidate any previous knowledge of its velocity. He showed that it is impossible ever to find out, at the same time, both the position and the momentum of a particle with unlimited accuracy. Certainly one of the characteristic properties of a particle is that it be somewhere and that it move from one place to another by passing through a series of intermediate points. But Heisenberg's analysis indicated that such a statement has really very little significance, since it is never possible to know where a particle is at the same time that one knows how rapidly it is moving. The early simple ideas of a particle have begun to lose much of their precision, and the question may be raised whether such a thing is properly a particle at all.

In spite of, or perhaps because of its obscuration of the simple picture of particles, quantum mechanics has led to the satisfactory analysis of a great many different phenomena. It has provided a theoretical understanding of the properties of molecules, of the nature of the light emitted by atomic systems, and has permitted a beginning of a detailed understanding of the physical properties of solids, in terms of the electrons and atomic nuclei of which they are composed. So successful has it been in this respect, that problems such as accounting for the magnetic and electric behavior of metals are often considered to require only the proper application of known principles to solve.

Quantum mechanics has also permitted a satisfactory attack on many problems of nuclear physics, although it is clear that in the realm of nuclear physics, quantum mechanics alone is insufficient to solve many of the puzzles that have been discovered.

Quantum mechanics has also been of interest to philosophers, although they seem not to have been so active with it as with relativity. But it has produced every bit as much, if not more, of a revolution in our ordinary methods of thinking. In spite of the fact that most physicists have a traditional mistrust of philosophy, the twentieth-century developments in the concepts with which physics deals have almost forced some consideration by physicists of problems which were formerly regarded as belonging exclusively to the domain of philosophers.

One of the problems which has occupied the minds of philosophers is concerned with the nature of existence. Does there exist a material world, and can anything be learned about it? On the whole, physicists have taken a very naïve view of such matters. To my mind it has been this fact, this naïveté of viewpoint, which has contributed much to the successful development of physics.

When physics entered the experimental phase about the time of Galileo, troublesome questions of existence were ignored, and it was assumed, often without adequate consideration, to be sure, that there existed an outside material world which could be contacted through the senses. It was also assumed that these sense contacts could be reliably interpreted, and that an observer could actually learn the nature of this outside world through his observations.

And at the present time, largely without a formal organization of their thoughts, most physicists regard the world as made up of two kinds of things: (a) Physicists and, if pressed, other persons or potential physicists will be included in this select group also; and (b) the outside world which can be studied and discovered.

Because of the simplicity of the physicist's attitude, the difficulties in his dualism were not at first troublesome. For instance, the question how a sensation got from the obviously material body of the observer to the obviously nonmaterial observer himself, was not a troublesome question; everyone could see that the sensation did get across the boundary line, and so what more was there to be said about it? Nevertheless, a little consideration of this problem makes it very formidable. Although it may seem quite clear that there are two kinds of things, observers and the material world which is observed, a little consideration shows that not only the nature of the boundary between the two but even the location of this boundary is obscure.

As far as the results of physical science are concerned, it seems possible to put this boundary at almost any desired point. The observer can, with apparent consistency, include any desired amount of this outside world in what he may wish to call merely extensions of his senses. The part remaining beyond appears to follow the laws of physics in a perfectly satisfactory manner, and there exists in the laws of physics no reason for assuming the boundary to be one place rather than another.

Consider, for example, that I wish to observe this desk. I am on one end, and the desk is on the other end, of a chain of interactions. Where shall I draw the line between myself and the thing observed? In the first place, I can draw the line at the surface of the desk, and say that the scattering of light from the surface constitutes the act of observing the desk. I can say that the source of light, the light itself, and all the mechanism necessary for the perception of the light, is part of me, is an extension of my sense organs. On the other hand, I could also say that what I really observe is the light, which strikes the retina of my eye, that this is the point at which the observation really takes place, and that here must be drawn the line dividing the observed things from myself. But I can go still further, and say that the action of the light on the retina is a purely physical process, which can be described by known laws, and that the dividing line must be placed at a point at which the nerve impulse reaches the brain. The fact that none of these places seems satisfactory might suggest that there should be no dividing line at all, except for the uncomplicated feeling that there must be made some such division. No one of these interactions shows any characteristics which distinguish it from the others, as long as only the classical physics is used; but one of the contributions which has come from the interpretation of the modern quantum mechanics is the recognition of the fact that, although the interaction which may be selected as the dividing line between observer and object is entirely arbitrary, and may be at any desired point, it is nevertheless necessary to put such a dividing line at some point, and to treat the interaction at this point in a unique fashion. To make clear the nature of this difference, may I outline briefly the method used in the quantum mechanical description of the behavior of an isolated part of the world.

According to the present theory, the state of a mechanical system is described or represented by a mathematical symbol, which I shall call the wave function. This symbol carries all the information which can be known about the system in the particular state in which it is. This symbol changes with the time, in accordance with a differential equation, known as the Schroedinger equation, in a perfectly definite way. If the state of the system is known at one time, it can be predicted for any future time, by means of this equation. In case the system is composed of two or more parts, the interactions and mutual influences of these parts are entirely described by this equation. But now suppose I want to examine the system. Suppose I want to see if everything is going on according to the rules, and for this purpose I want to make a measurement of some quantity. As soon as I touch the system with a measuring instrument, as soon as I make any kind of contact with it sufficiently vigorous to learn anything about it, the symbolic wave function explodes in my face. The interaction between the system and myself, in the form of a measuring instrument, cannot be made gentle enough to leave the system undisturbed, and at the same time strong enough to give me some information. This interaction cannot be described by the same Schroedinger equation which described the behavior of the system as long as I did not touch it. Thus there is a distinct difference in the treatment accorded interactions which take place within an isolated system itself, and those which take place with the observer, or the observer's extended senses, in the form of measuring instruments. It is just possible that this difference in treatment may be of significance beyond the regions in which it has thus far been applied.

One of the subjects often discussed in connection with the implications of physics is the problem of causality. One hears frequently about the principle, the law, and, more recently, about the disappearance of causality from the world of science. The principal difficulty with this subject seems to be to find out what one is really talking about.

Apparently one of the essential elements of causality is that events shall have some connection in time, that the occurrence of a certain event now is necessarily followed by a certain other event at some later time. I do not mean that this is all that causality implies, but this seems to be at least one thing.

This causality, this uniform development in time, has been assigned in quantum mechanics to the symbolic wave function. The symbolic wave function, and its law of change, carry within themselves the usual conservation laws, such as the conservation of energy, the conservation of momentum, and the conservation of angular momentum. On these conservation laws rests the usual idea of determinism. The symbolic wave function does not carry, however, a detailed space-time description of the motions of the particles of which the mechanical system is composed. The wave function is quite an abstract thing. It cannot be observed directly, and its connection with observations to be made on the system is, in general, only statistical. Thus it is true that the present mechanics does not permit an exact prediction of the result of a measurement to be performed tomorrow. It permits only statistical or probability predictions to be made in most cases.

Does this mean that there is no causality in physics? That still depends on what you mean by causality, upon what you want causality to do for you. To many persons the term "causality" is associated with the ideas of determinism and freewill, and the significance to be attached to the problem is because of its connection with ideas of moral responsibility.

At the time of the rapid development of Newtonian mechanics, and its phenomenal success in describing and predicting the motions of the members of the solar system, there grew up the belief that all problems were to be solved by such essentially mechanical means. In particular, it was concluded that our conscious mental processes were to be determined, and described, in terms of motions of atoms in our brains. Although such a conclusion is clearly at variance with the simple dualism in terms of which physicists normally think, there were many persons who believed it to be a direct consequence of the thinking of physicists. The discovery of the statistical element in the predictions of quantum mechanics was seized upon by some as a means of escape from these unpleasant conclusions. It was suggested that although natural laws operate in all phenomena, they are not to be regarded as determinative, but merely as restrictive. Inside the range permitted by the statistical laws, freewill might be supposed to act.

In spite of this suggestion, I think it is now agreed by most physicists who have considered the matter, that the conclusions from Newtonian mechanics to a materialistic determinism in phenomena of consciousness, as well as the conclusion from quantum mechanics to a possible freedom of will, are entirely without any justification.

Nevertheless, there has grown up under the influence of

Bohr a recognition that certain aspects of the methods of quantum mechanics may provide a point of view useful in problems of this kind. The machinery of quantum mechanics provides for certain conservation laws, but does not at the same time provide a detailed space-time description of events. There are in the problems of atomic physics two complementary but mutually exclusive aspects. Both of them are necessary to a complete description of the phenomena, but neither of them is adequate by itself. For instance, an electron is found to behave under certain circumstances as a wave, and under other circumstances to appear to have a clearly localized position, as if it were a small particle. The achievement of the theory is in renouncing any attempt to describe one of these aspects in terms of the other, or to establish any detailed connection between them, and in the recognition of this complementarity as fundamental. Certainly waves and particles are not the same thing; in fact, they are mutually exclusive things, and the recognition that, in spite of this, an electron has properties of both kinds, is a real change in mode of thinking. In some such way one might imagine that problems of consciousness may have two complementary aspects. One of these aspects might suitably be described by such words as "freedom of choice," while the other might be described in terms of physical or chemical reactions. The progress in understanding would come with the recognition that one of these descriptions does not exclude the other, but that they represent entirely different aspects of the problem. This rather surprising point of view, which has been forced upon us by the results of actual experience, may be one of the major contributions which physics has to make to philosophy.

In these three groups of ideas in the physics of the twentieth century it is interesting to note a somewhat changed attitude toward experiment and reasoning from that previously in vogue. Galileo has often been referred to as "the father of modern science," the real initiator of the scientific method. In a recent lecture, Professor Cohen of Harvard has given particular attention to some of Galileo's work, in commenting on the doubtful accuracy of many current beliefs regarding the history of science. He pointed out that Galileo was not a strict empiricist, that although he made observations and performed rough experiments, his conclusions were based less on the experiments than on his conviction of their innate consistency. Galileo puts in the mouth of one of his characters the following words, "I, without experiment, am certain that the effect will follow as I tell you, because it is necessary that it should."

And in a later century, in an age which has been called the "age of reason," it is natural to find a great reliance placed on rational thinking, and such thinking must be based ultimately on intuition. This insistence upon understanding as a function of science persisted through the nineteenth century. Since mechanics was then such a well-developed science, a phenomenon was regarded as satisfactorily explained when a mechanical model could be made for it.

Possibly the great impression which Einstein made upon his world was due to the way in which he accepted the results of experiment as being more authoritative than any apparent irrationality which they seemed to display. Einstein took the results of the Michelson-Morley experiment, realizing that they were incomprehensible, and undertook to draw conclusions from them. The conclusions, as you all know, have been of far-reaching significance. In the same way, the founders of quantum mechanics took the results of experiment and undertook to formulate them into a system. We are not able to make a rational picture of the behavior of atomic particles. Sometimes they are particles. Sometimes they are waves. But it is the achievement of theoretical physics that it has been possible to set up rules which permit an unambiguous prediction of the ways in which particles will behave under given circumstances. Through many trials and tribulations, physicists have come to accept experimental results as the one and only authority to which they can appeal.

Perhaps the only real point here is that the experiments on which modern physics relies are much more complex than the ordinary experience common to everyone. This fact, however, does emphasize the important principle that physical laws, theories, and ideas cannot be extrapolated beyond the regions in which they are shown to be valid. In this state of mind a thoughtful physicist is inclined to be humble. No longer can he follow Descartes as he said, "Give me extension and motion and I will construct the world." More likely is he to be impressed by Hamlet's "There are more things in heaven and earth, Horatio, than are dreamt of in your philosophy."

It is largely within the past decade that physics has moved to its present position of prominence. It is at the same time feared as a destroyer of civilization and regarded as its only hope. For the physicists who have lived through this period, the change has been disturbing in many ways. No longer are they individuals, working largely by themselves, or with small groups of students. Even in the universities, physics departments have become more formal organizations, with sections and leaders, with organization charts, with business and budget departments, and with secretaries busily making carbon copies of many things. Financial support is available in unprecedented amounts, and the general philosophy is strongly reminiscent of

Francis Bacon and his plans for a concerted human attack on the mysteries of the universe.

But in the midst of all the activity, a significant number of physicists are inclined to question where this effort is leading. Is this concerted attack really the kind to be effective, or is there a danger of losing in the confusion something essential and vital?

The philosophy of the physicist before the twentieth century was colored by the fact that the world at large did not impinge too deeply on his life. The physicist of the last century seemed relatively impervious to outer environments, because he carried his own inner environment with him wherever he went. It is reasonably clear today that the problems of defense and survival have made this detachment an impossibility. In the exigencies of the moment, however, let us not forget that the door to the inner environment of detachment must be left ajar. The wars of '17 and '41 have stimulated enormously the application of many physical principles. The ideas themselves, however, were born in the quiet, somewhat desultory, atmosphere of college laboratories. Man apparently can manufacture in concert, but he thinks alone. Production can be organized and planned, but the only climate in which physical research can breathe is the climate of freedom.

We find that precious climate beginning to fog in from many angles. Because it is an absolute we must have, we had better gird up our loins and do battle for it. Mr. DeVoto's words in the *Atlantic* for February 1951 are even more applicable to science than to letters in their delineation of the choices confronting us. "Freedom," he points out, "lays on anyone a heavy burden of obligation, it is the heaviest of all burdens. . . . No matter how inscrutable the future, the axiom at the basis of American experience is that freedom counts most of all."

We have seen men who have "fled from freedom to the absolutism that relieves a man of responsibility and his mind of obligation, to do the mind's work." We have seen the foes of democratic processes "make use of the American belief in freedom, as an instrument for the destruction of freedom." . . . The original problem of individual responsibility "has not changed since Jefferson wrote, 'I have sworn, upon the altar of Almighty God, eternal hostility against every form of tyranny over the mind of man.' It has not changed since mankind began to work out the conception of free men in a free society. The enemy is still the same: absolutism, authoritarianism, dictatorship, tyranny, whatever threatens freedom. And the burden that freedom lays on the human mind and soul has not decreased; it has grown heavier. The strength of every honest man is needed in its support."

I do not know how many physicists I may speak for in the matter of the nation's defense, so I shall speak only personally to say that we need have no real despair over winning the battle of arms, and the battle of scientific supremacy. We can, I firmly believe, do that. But any thoughtful scientist, and any thoughtful citizen of this democracy does often despair of winning the battle of men's minds. Lies make quicker allies than truth. One can only hope that truth can make more lasting friends. There are a billion people listening in the farthest corners of the earth tonight to two quite distinct voices. The first one says, "You are poor and hungry, because the rich have held you down—revolt!" The second says, "Knowledge is power. Learn how to make the earth yield you a good life. Learn the secrets of its minerals. Learn to put the elements back into the earth that will enrich it, and enrich your children. It is not force of arms you need, but hydraulic dams, and flood control, and hybrid corn." And it is the will to live in freedom in a just world, on which we must make our stand. Let us not, then, let the temporary clash of arms drown out the small voice of man's longing to know the world's truths, and to build on them steadily for a sounder tomorrow.

W. ALBERT NOYES, JR.

The Varied Uses of Chemistry

The word "chemistry" itself is by no means easy to define in such a way that it would include all that so-called chemists think they do, and the word "chemist" is variously used by the lay public to convey everything from a veiled insult to open admiration. If chemistry were defined, as it has been classically, as the study of matter and of its transformations, it would include nearly the whole of science and of metaphysics from the prosaic to the sublime. Who can deny that atomic energy, genetics, cosmetics, fuel, and fertilizer, as well as much of astronomy, would be properly classified as chemistry by this broad definition? This might simplify our university structure, much to the joy or chagrin of deans and college presidents depending on the effect on endowment campaigns, for one could abolish nearly all scientific departments except chemistry. As for the chemist, he may be variously a druggist or pharmacist, the high school graduate who washes the glassware and carries out routine analyses, the president or vice-president of a corporation, or even a dean or college president who would no longer be able to pour a liquid from one beaker to another.

It may be difficult either to define a physician or to limit by any single definition the field of medicine, but the task as regards chemistry and chemists is truly hopeless. He must rarely be licensed, he need not have any kind of a college degree to announce to the public that he is a qualified chemist, and unfortunately he is not bound by any kind of oath or well-specified code of ethics. When some chap explains to you that he is a chemist, you may justifiably raise your eyebrows if he has only completed the eighth grade, but circumstances would have to be quite unusual for you to be able to take any legal steps against him.

I have therefore given myself considerable latitude in the subject I have chosen, and I doubt if anyone can criticize this talk for having wandered away from its central theme. Nevertheless, we shall try to follow the allegedly scientific method of proceeding logically from certain premises to certain conclusions.

It would be useless to try, even if one could succeed, to gather into the field of mathematics all branches of science which use arithmetic, or algebra, or calculus, or even less common fields such as vector analysis and the theory of numbers. In the same way chemistry is of enormous utility to many sciences from medicine to biology, to geology, to agriculture, to physics, and even to psychology. A problem should not necessarily be classified as chemical because some aspect of chemistry must be used in its solution. The chemist may be called upon to make all sorts of compounds or to develop methods for determination of either the purity or the composition of substances under conditions such that the science of chemistry itself will be advanced little or not at all. The real advance comes in some other field for which the chemist merely performs a service. How much is engineering advanced by building another bridge quite similar to, even though not identical with, hundreds of others already built? Certainly hard work may be needed, and often much ingenuity must be used, for each bridge has some new difficulties which must be overcome during its construction. The real question is whether some entirely new principle of bridge building has been developed or whether it has been necessary to provide new materials of unusual properties.

Thus I wish to distinguish clearly between those advances in basic science which might in the future, probably distant future, modify our fundamental thinking about chemistry, and those advances of perhaps equal but more immediate importance which advance chemistry little as a science.

There have been a few discoveries without which chemistry could not possibly be what it is today. We could not in a brief time list them all, but by way of illustration we will name a few. First and perhaps foremost is the law of conservation of matter, fortunately not upset by the reversible transformation of matter into energy until chemistry had achieved a solid foundation. The indestructibility of matter, which implies that it may only be transformed, is a concept truly basic to chemistry. To put it in more modern terms, atoms may be separated from their present unions and combined with new ones, but they continue an uninterrupted and eternal existence until either man or nature provides some truly cataclysmic phenomenon.

Another important discovery was the law of multiple and combining weights. Thus either one gram or two grams of hydrogen will combine with sixteen grams of oxygen to form either hydrogen peroxide or water. After long and careful experimentation we are firmly convinced that we have here a law of nature and that we shall not someday find a compound with one and a half grams of hydrogen combined with sixteen grams of oxygen. It is this law which leads us to believe in atoms truly indivisible except under extraordinary circumstances. Thus the atom of oxygen must weigh either 4, or 8, or 16, or 32 times as much as the hydrogen atom. We will not bore you with trying to explain why we write the formula of water as H_2O and that of hydrogen peroxide as H_2O_2.

Another great and fundamental discovery, which made possible the systemization of organic chemistry, was the so-called tetrahedral carbon atom. Perhaps this should be called a theory rather than a law, but it explains so many things which could be explained only imperfectly and artificially by any other means that we consider the tetrahedral carbon atom to be a fact, and no longer permit it to be treated with levity. Briefly it may be described thus: Imagine a tetrahedron standing on one of its triangular bases with the tip up in the air painted black. Now paint the three bottom corners red, yellow, and green as we go around them clockwise. Bring up a mirror and look at the image of the tetrahedron. The base of the image will appear to have the colors in the order red, green, and yellow, as we go around it clockwise. In other words, we can use the same four colors on the corners of the tetrahedron and arrive at two results, one of which looks like the mirror image of the other.

This may all appear like child's play and may make you wonder whether chemistry should not be in the kindergarten rather than in the graduate school. Nevertheless, a simple idea like this required a lot of imagination, and it does explain why two compounds, with the same composition and even almost the same structure, may yet differ in certain properties.

Another discovery of great fundamental importance was made much more recently. I refer to the discovery of nuclear fission. Undoubtedly the claim that this was a basic discovery in chemistry will arouse the ire of our friends the physicists, but as a matter of fact the person who received the Nobel Prize for this discovery, Otto Hahn, is a chemist, and but for the techniques commonly used by chemists this discovery could not have been made. This great achievement, which paved the way for atomic bombs and perhaps for atomic power, opened new horizons for all fields of science.

These illustrations have been chosen deliberately because they have all affected the fundamental trend of science. It is always difficult to predict whether or not some new and original approach will open up broad horizons. The environment must be right for a really important new idea to be recognized and developed. Many great ideas undoubtedly have not been recognized as such, either because they were too far in advance of scientific thinking at the time they were proposed, or because scientists were too preoccupied with other matters to give them proper attention.

What is the environment which is necessary? Indeed, what do we mean by basic research? In a day when the gadgetry of science costs huge sums and when we have large teams dealing with narrow problems, we may well ask ourselves whether increase in physical plant and employment of more scientists and technicians will change our basic concepts more rapidly than in the past. Certainly a lot is being done under the name of fundamental research which could not by any stretch of the imagination be called fundamental. Nevertheless, there is some tendency among scientists, as there is in government, to measure a man's worth by the number of persons under him on an organization chart. The top man, who allegedly arrived at his position because of scientific ability, is kept so busy raising money and accounting for it, lecturing and running around, that he has little time to think about the research for which he raises the money and which forms the subject of his lectures.

The picture may not be quite so bad as I have painted it. A French friend of mine once told me that in his opinion the success of American science is due to our ability to make effective use of mediocre talent. The accumulation of much information by a lot of routine work is prerequisite to the solution of some problems. I am not referring to the so-called Edisonian method where one tries every bottle on the shelf without any guiding thought. The only system in such an approach is alphabetical if the bottles are arranged in alphabetical order. I refer rather to very complex problems for which several solutions appear possible until many facts are at hand.

We can cite two examples. One lies in what was formerly a branch of physics, but has now passed over mainly to chemistry because the physicists have become interested in other matters. I refer to spectroscopy. Here we are dealing with a field where precise measurements can and have been made. The beginnings of the interpretation began about the time of World War I, when Bohr first introduced the idea that atoms can exist only in certain states of energy. Scientists, perhaps more than most others, are unorthodox and uncontrollable individualists. This freedom of thought and action, which they so cherish for themselves, they found was denied to atoms which have the alternative of being only on certain floors of an edifice, but never in between. They can absorb energy and jump up a floor, or they can give out energy in the form of light if they fall from a higher to a lower floor. This idea was susceptible of some simple tests. Suppose an atom falls from the eighth floor to the fourth floor, and another atom from the eighth to the fifth. The energy for the first kind of light, less the energy for the second, should be equal to the energy given out when an atom falls from the fifth to the fourth floor. To put this statement in somewhat more scientific form, the frequency of a spectral line is often equal to the difference in frequency of two other lines.

It is always dangerous in science to make quantitative predictions, because one might be proved to be wrong. How much more comfortable to live a placid, even if uninspiring existence? Bohr's theory did make some quantitative predictions, and they were found to be correct. It soon developed, however, that the floors of the atomic edifice were not all alike and that under some conditions they could be moved a little. Thus it became necessary to describe the edifice in precise terms and to account for these minor changes. The theories of the atomic edifice have grown and been given substance, so that today the picture is much more complex than it was originally. The filling in of all the information necessary to test these complex ideas has been a slow and laborious process. It could not have been done by one man. Thus the theory has been amplified bit by bit, mostly by additions which were not in themselves striking, and experimental verification has sometimes preceded, sometimes followed, the theory. Spectroscopy has very largely got out of the realm of basic research. Probably really new ideas will be forthcoming, but most workers in the field are merely putting bricks on a wall for which the pattern is already set.

The other example I wish to discuss lies in a field of

great interest to me: the speed of chemical change. The chemist not only wants to know what he will get, but how fast he will get it, and how much. The obvious connection of these problems to dollars and cents need not be discussed. Speed is never a thing easily measured, particularly if it is changing and depends on many factors. Here we encounter one of the most difficult problems in chemistry. In a general way we can talk learnedly about the factors which determine whether we must attain the age of Methuselah before something will happen, or whether we shall have an explosion which might destroy both apparatus and experimenter. In reality the basic parts of this problem have not been solved. This is a field begging for men with ideas as well as for teams of men with patience and the capacity for precise experimentation. Teamwork will not provide the answer, but will furnish the basis for some great thinking. Since the problem is so difficult and since present ideas are very crude, one must hope that a truly fine mind will focus its attention on this problem.

This brings us to one of our main points concerning basic science: the real need for more theorizing and more correlating. Few of the basic ideas of science have originated in this country. Often they have come from European scientists with poor laboratories and meager salaries. There is something they have had, however, which we Americans do not have, perhaps because we do not appear to want it. That is leisure. One may be able to build apparatus a few minutes at a time, or do administrative work on the fly, but the solution of the really fundamental problems of science demands thought whose train must not be interrupted too frequently by telephones and visitors. The American proclivity for red tape and disoriented activity does not provide the proper atmosphere for the really great advances in any field of learning.

I wish I could tell you what the great basic problems of chemistry are and how we should try to solve them. This I am unable to do, partly because of my own shortcomings, and partly because a problem cannot always be recognized until some experiment has been performed which is inexplicable by current thinking. The difference between what we find and what we think we ought to find often leads to great discoveries.

It may be worth while, however, to point out some of the really basic problems of chemistry that are now begging for solution. Most of these problems arise from the fact that chemistry usually deals with complex systems involving large numbers of atoms. Not often is it possible to devise experiments which depend only on a single atom or particle. This has been done to some extent in nuclear physics, but chemical reactions rarely lend themselves to such a simple approach. The statistical approach must therefore be used, unless one wishes to deal with crude approximations and somewhat meaningless generalities. We may illustrate by an analogy. It is unwise to predict how long a particular person will live. He might be killed by an automobile tomorrow, or he might live to be a hundred. However, it is possible with the present statistics concerning death by accident, by disease, and by old age to predict with some certainty what fraction of the persons now fifty years of age will live to be sixty. The prediction might be wide of the mark if atomic bombs begin to fall or if some new disease were to decimate the population.

Thus the chemist by the use of statistics can do something about predicting behavior as long as nothing extraordinary is happening, but his approach is indeed hesitant if he has to deal with the time factor. The application of the statistical approach is extraordinarily difficult when all the factors bearing on a problem are dimly understood.

At times one is led to wonder whether the use of rigorous mathematics to solve the complex problems of chemistry will ever lead to fruitful results. The use of sufficient approximations will vitiate the best mathematics, and one can be fooled by a beautiful derivation which bears no relationship to reality. One must not counsel despair, however, for the goal must be the correlation of data by mathematical expressions. Otherwise the human mind will find it impossible to assimilate and to use the vast array of facts at its disposal.

Thus we can state that the chemist must eventually use the approach of the mathematical physicist, realizing that his problems are very complex and that he cannot expect to use that approach in a simple and direct way. The fundamental bases of the statistical approach need reëxamination and amplification.

Few minds are capable of attacking the basic problems of chemistry by the method indicated, and we certainly should not encourage the incompetent to try anything along these lines. Perhaps the world would be better off if we had less unwarranted theorizing, but chemistry certainly needs to attract more minds capable of rigorous mathematical analysis.

We do not wish to imply, however, that some basic problems are insoluble except by the mathematically minded. There are many such problems in biology awaiting the chemical approach for which the usefulness of the exact physical approach is not apparent. Again we can only cite examples without hope of covering the field.

The sulfa drugs, penicillin, streptomycin, and indeed the whole field of chemotherapy, present problems to the chemist of unlimited scope. Not only must these compounds be synthesized and their structures determined, but the way they act must be understood. In a real sense this field has developed in a very empirical fashion. Certain substances are beneficial for the treatment of certain diseases, but reasoning by crude analogy is about as far as one can go in predicting what new substances might be better than those already known. There is too little basic information available to warrant the use of real theoreticians. One has only vague ideas of what to measure quantitatively as a basis for future theorizing, but certainly much laborious basic work must be done to find the pertinent variables. The problem is similar to the old familiar one in industry and in the Armed Services, namely, that of establishing criteria which may be used to determine whether something is good and if so just how good. An antiknock compound must be added to gasoline and then tried in a motor. An antibiotic must be tried first on sick animals and then on human beings. In neither case is the approach basically scientific because one only observes the end result, i.e., good behavior of an engine or cure of a patient, without understanding why. Only by sheer good luck could either the best antiknock or the best antibiotic be found by any such method.

In the field of cancer we have a great opportunity for the chemist. Much useful information is being accumulated, and there has been real progress in recognizing early symptoms and in treating patients, but the basic problem of why certain cells without obvious reason become abnormal and start on a wild career of growth remains unsolved. Useful clues are apparent, but one of the difficulties has been, and still is, to learn what to measure before a patient develops cancer so as to be able to warn him of his danger from this disease. Systematic studies are being made, and the data being accumulated are so vast that it is difficult to see the forest because of the trees. Possibly, although not probably, the answer is staring us in the face waiting to be recognized.

The great problems of biology, such as heredity and cure of disease, are basically chemical in nature, but before the talents of the chemist can be used effectively there must be amassed a large amount of more or less empirical information. New techniques must also be developed for studying individual parts of cells. The scientist who enters this field must be well versed in many disciplines. Teams of workers with varied backgrounds are essential, but the guiding genius must be a man of great imagination as well as of varied training. Such men are rare.

We have dealt so far with some basic problems of chemistry as well as some basic problems in other fields whose solution will depend ultimately on chemistry. What is the future of chemistry in industry and in government?

The discussion of these matters will be colored necessarily by the present emergency, which may be of long duration. Many items will be scarce, and many chemists will be employed in devising substitutes. Occasionally this work will lead to new and improved products of use in peacetime, but more often this work will have little long-range value. Many other chemists will be engaged in development and production of devices to be used by the Armed Services. When one remembers that even before the present emergency about half the money spent on research in the United States was derived from the budget of the Department of Defense, and that this fraction is bound to increase, one can see to what extent science has been diverted to war matters. Of course some of this money advances pure science, and some more will go into applied science of permanent value, but most of it can be justified only for defense of the country.

It is unsafe to predict the future on the basis of the past. During World War II, training in chemistry fitted persons to engage in a large number of miscellaneous activities not purely chemical in character. An intelligent person well grounded in the fundamentals of science may be adapted to very diverse activities. Thus many chemists for the next few years will not be practicing the profession of chemistry, even neglecting those who are drafted. It was estimated that during World War II less than forty per cent of the chemists in the Armed Services were placed where they made any use whatsoever of their professional training.

The manpower problem is very complex. Are chemists in really short supply when the situation is viewed from the standpoint of the national interest? Anyone who has tried to employ a qualified chemist during the past few months would agree that employable chemists, particularly good ones, are hard to find. Chemists, however, are not exempt from the frailties of the rest of the human race. Corporations and academic institutions alike would prefer to keep their groups intact, partly because if they are able to do so, they might have a real advantage over competitors at the close of the emergency. Moreover, it is not thrilling to leave a pleasant home and interesting work with which one is familiar and move either to Washington or to some out-of-the-way arsenal or an overcrowded industrial city.

Let me hasten to add that I believe chemists to be as patriotic as any other class of Americans, but they need guidance in the present confused situation. A great deal of useless work is being performed under the guise of defense research. This is obvious even to the casual observer, and to those on the inside the situation appears to be serious. Is it important to continue work on cancer and to develop new products which could be important in the post-emergency economy, or is it better to drop these pursuits and do research work on new war gases and explosives? The decision confronting the individual chemist is a most difficult one. Gas warfare was not used during World War II, and new explosives, even after their synthesis in the laboratory, must follow a long and arduous trail before they can be mass-produced and used in combat. What guarantee against complete wastage of effort is there if a chemist changes to a new position?

Obviously no ironclad guarantee is possible, and any bureaucracy established to make decisions on these matters might make decisions which would hamper the war effort seriously. Some intelligent guidance is possible, however, but such guidance can be given only by a competent organization whose members have a broad view of the needs of the country. No such organization exists today. The poor chemist is being approached by many defense agencies to undertake things which he only dimly understands. He must decide with insufficient background what is most important. He would be comforted if told by competent authority that his present work is more important than much so-called defense research.

We now come to a very controversial field about which I have some convictions. My opinions may be worthless, but I am going to put them forward. They are based on the belief that we shall have a long state of preparedness for war even if global war does not come.

It cannot be said that we were really short of chemists during World War II. Even at the close of the war there were still many chemists engaged in activities of little importance either from a scientific or a nationalistic viewpoint. It is true that certain types were scarce, and it is also true that we were beginning to be short of young chemists.

Had the war lasted another five years, we should have had some serious manpower problems in science. Not only do the new ideas very frequently come from young scientists, but a good man who has had a degree for a few years should become the leader of a group. This is impossible if the output of chemists is seriously curtailed.

The first point is, therefore, obvious. We must not stop training scientists for any extended period of time. We do not need to train as many as we did during the immediate postwar years, but we should never let the level fall as low as it did during World War II. Moreover, the choice of those we train must be based on intellectual capacity rather than on either physical disability or the ability of some professor to obtain an alleged war contract. The situation needs to be faced honestly for the good of the country. A student of great ability must not be criticized for not being in the Armed Services. He should be prevented legally from being in the Armed Services for any extended period of time.

The kind of work to be done by chemists for the Armed Services is worthy of brief discussion. It is unsafe to generalize too broadly, but the time interval between test tube and use for most items is usually several years, even in wartime when no need exists to make profits and practice economy. Thus little laboratory work can have a real effect on the way wars are fought except on a long range basis. Nevertheless, scientists do have much to do. Occasionally they will develop something radically new, such as radar or A-bombs, which will change the way wars are fought. More often there will be improvements in existing items which will lead to increased effectiveness.

The modern war is fought with a vast array of complicated devices. Not only must these devices be capable of use by troops, they must be reliable, and they must be used effectively. Tests must be devised which will enable predictions to be made of effectiveness, and doctrines for use must be prepared so that desired results can be achieved in combat. The problems will vary in complexity from the determination of the usefulness of gloves in a Korean winter to the type of bombing which should be practiced by the Air Force against an enemy dispersed over a wide area and protected only by slit trenches. It is imperative for a commanding officer to be able to predict with some certainty how much resistance his troops will meet after *x* tons of bombs have been dropped on *y* square miles.

These questions are difficult enough to answer concerning some device which has already been used in combat and about which information has been obtained by military intelligence or through aerial surveys. They are extraordinarily difficult for some new and untried weapon.

Soundly trained scientists with vivid imaginations can be of great service in this broad area of practical testing and evaluation of field data. New and dreadful methods of waging war are often described by scientists themselves, and these make scare headlines in the public press. Will these headlines merely lead to unjustified panic in our cities at the start of a war, or are these weapons really effective? Scientists too often lose their scientific objectivity when they step outside their own little provinces, and here we have an example of a time when they should keep quiet until they are sure of what they are saying. A study of history will reveal that the human race has great resilience and resourcefulness, and its destruction is more difficult than it might appear to be at first blush.

The Armed Services need more scientists with broad backgrounds and a flair for the practical. They can be of inestimable value not only in developing new gadgets but also in helping to decide how old ones can be used effectively. The officer at Bunker Hill who ordered his troops not to fire until they could see the whites of their enemies' eyes had a practical, scientific sense even if he did not have a Ph.D.

I would like to make a plea for more scientists to step out of the laboratory and forget their own peculiar interests. The scientific mind is needed in areas for which it was not specifically trained in peacetime.

So far we have treated of basic science and have skated around the fringe of some military problems. Our treatment of the latter has been circumscribed by security regulations. What of chemistry in industry? This broad topic could not possibly be treated in detail in a short time, even if I were the person to do it.

This country has been expending its natural resources at a stupendous rate. Those who feel that the American system of industrial enterprise is solely responsible for our high standard of living should realize that few other nations have had within their own borders such a vast array of mineral and agricultural resources. Would it be possible, for example, to have as many automobiles per capita in Western Europe as we do in the United States? Not only would the density of traffic be appalling, but all of the gasoline would have to be imported unless it were made from coal. Even coal is not abundant in all of these countries. Imports can be paid for only by exports, and what could Europe sell in sufficient quantity to make a really large-scale use of automobiles possible?

There are some essential materials we do not produce which are not produced in adequate quantity anywhere in this hemisphere. There are others which would be in short supply if all countries of this hemisphere were to attain our level of industrial output. These facts should be remembered by those who would have us withdraw to our own shores and forget our allies overseas.

It appears inevitable that chemists will be called upon to furnish new products and new ways of getting old ones. As rich ores are depleted, ways must be found to use low-grade ores. Many different products are now made from petroleum and natural gas. The public probably does not appreciate fully the size of the chemical industry based on these natural resources. Gasoline and lubricating oils are still the main products, but textiles, plastics, synthetic rubber, pharmaceuticals, dyes, and a host of chemicals essential to industry are made from crude oil and natural gas. The end of petroleum reserves is not in sight, but it will be necessary someday either to change our economy radically or to shift to coal as a source for many of these things. The chemist will play a vital part in this development when it comes.

In the last analysis, however, a high standard of living is dependent on cheap power. A rough index of standard of living could be obtained by dividing available horsepower by population. Such a yardstick will show the United States to lead the world by a substantial margin.

What are the sources of power? They may be counted on the fingers of the two hands: (a) manpower, which in turn depends on food; (b) animal power, again dependent on agriculture; (c) coal; (d) petroleum and natural gas; (e) sunlight; (f) water power; (g) subterranean heat as used in Italy and one or two other countries; (h) tides; (i) wind; and (j) atomic power.

We may say at once that manpower and animal power alone could not support an industrial economy. By use of slave labor, as in ancient times, a small fraction of the population might enjoy a comfortable standard of living, but this fraction could never be large. Water power is more useful, particularly in certain mountainous countries where the population density is not too high, but it is insufficient in large areas of the world to support a modern industrial economy. Subterranean heat is available in few localities. The wind and tides could be harnessed only at great expense, and would be inadequate of and by themselves. Thus we reduce our list to petroleum and natural gas, coal, sunlight, and atomic power.

Petroleum and natural gas are abundant, but not too widely distributed over the earth. Improvement in the economy of other nations may make it difficult for us to have all we want. Plans should be made now for use of other sources of power. Fortunately we have large reserves of coal. Coal mining is not a pleasant occupation, and more research is needed on how to use coal more effectively and on how to convert it to liquid fuel. Liquid fuels can be transported more easily and burned more efficiently than solids. Processes are known for converting coal to liquid fuels, but much needs to be done before they could be used on a large scale.

Sunlight is theoretically a potentially great source of power, but man has not devised a better process for storing this energy than nature has provided in the growth of plants. There are several leads which could be followed, but none looks too promising. Work on this problem is proceeding in a rather desultory fashion in several laboratories, but most of it centers around attempts to understand how plants use sunlight.

This brings us to atomic power. The dreams of the immediate postwar years have been replaced by hard realities in 1951. Perhaps atomic power for other than military purposes will not have the highest priority as long as we are making atomic bombs. In any case, there are difficult problems to solve, notably the obtaining of sufficient raw material and the disposal of wastes. Purely engineering difficulties are also considerable. At present it appears doubtful whether atomic power will replace other sources of power on a large scale in the foreseeable future. Nevertheless, here is a problem for the chemist and the engineer which will challenge his ingenuity and could be of immense benefit to the human race. We need an extended period of freedom from the anxiety of war if atomic power is to become a reality on a large scale.

When the problem of power is examined from a world rather than from a strictly nationalistic standpoint, it is probable that our sources of natural fuel are adequate to maintain a reasonably high level of economy for many years to come. Artificial barriers created by man prevent migrations of populations to areas adequately supplied with power, and prevent further the exportation of the products of labor to the extent necessary to permit purchase of fuel in heavily populated countries. As long as these barriers exist, the power problem alone will prevent the attainment of a high standard of living for a considerable fraction of the world's population. The solution demanded today is, therefore, the achievement of cheap power through materials which need not be imported continuously. It is well known that uranium ores are not widely distributed unless one begins to use very low-grade materials for which the cost of extraction of uranium would be very great. Nevertheless, the possibilities of atomic power are such that if these materials are used, the situation could be alleviated quite materially. Here is a real problem for the chemist, but one whose solution demands removal of some of the secrecy which now surrounds this entire area.

While the problem of power is really basic to the future of the human race if a uniformly high standard of living is to be achieved, the problem of food is equally important. The average American consumes food, expressed in calories per day, at a rate equaled in few other countries in the world. Perhaps the American diet provides more food than is necessary for a happy existence, and also the wastage in this country is stupendous. The amount of food we burn and throw away would provide more than a subsistence level for a large population. Perhaps the standard stated as desirable in our government publications is much higher than necessary, but if the world as a whole were to eat as much as Americans eat, it would need a food production perhaps 30 or 40 per cent greater than it now has. We now know much more than we did formerly about vitamins, amino acids, and other essential elements in the diet, so that it is possible theoretically to use food more effectively than it was ever used before. Also we know more about optimum methods for growing crops, so that yields per acre could be increased, although one must not forget that fertilizers must be plentiful unless the soil is to be depleted, and fertilizers can be made only from atmospheric nitrogen by use of considerable power, or from certain nitrates which occur in nature. The arid zone is also gradually encroaching on arable land, a process which must be stopped if the food supplies of the world are to be adequate. Animal sources of proteins are insufficient and must be supplemented by more use of sea food and of suitable plant proteins.

The problem of food and the problem of population control cannot be separated. The chemist may help solve the problem of food production, and he may aid in solving the conquest of disease, but sooner or later the present rate of growth of population will make it impossible to have the kind of standard of living that we as Americans have come to believe desirable.

All of these problems concerning power and food are intimately connected with the broad problem of education. In this country nearly every boy or girl can go through high school even if his or her mental equipment is modest, and even though he or she may come from a relatively poor family. An astonishingly large proportion of those who are graduated from high school go on to college, and of these a fair proportion is trained in the sciences. The total number of students in our universities is probably greater than in all of the rest of the world combined, although the figures in the Soviet Union are unknown and may make this statement erroneous. There are more members of the American Chemical Society than there are in all of the other chemical societies in the world by a considerable margin. The large amount of technological training received by our young people, combined with a great abundance of raw material, has placed the United States in a position which is unequaled both from the standpoint of agricultural and of industrial production. Other countries, even those with an abundance of raw materials, are less able to achieve this level of production because they do not have sufficient trained personnel. This problem has been recognized by the United Nations as well as by some of its specialized agencies, and was indicated quite clearly by President Truman in what is often referred to as his Point Four.

The training of scientists must be based upon sound primary and secondary education, and to achieve the latter, the basic philosophies of certain populations must be altered. In the Arab countries, for example, there is a large mass of poverty-stricken and illiterate people on which is imposed a superstructure of a relatively small merchant class and an extremely small, from a percentage standpoint, wealthy class. Higher education is confined almost entirely to these upper levels of society, and they have been brought up traditionally to believe that work with one's hands is degrading. If these countries desire to make use of their own natural resources, they must approach the problem by a series of slow and painful steps. It is first necessary to improve and to broaden the base of primary education to the end that a large fraction of these populations will begin to understand the place of science in modern life, and to realize that a career in science is just as important as a career in government or in law, even though it may not lead to the same position of power in a community.

Once the level of primary education has been raised, it would be possible to select a larger number of students who could be advanced through secondary school, and finally there would be more students available for university education. All too often our philanthropic organizations and governmental agencies tend to put the cart before the horse by establishing research institutions and institutions of higher learning before the broad base of the population has been educated to the point where such institutions can flourish. It must be recognized that any program such as the one which I have outlined would cost money, and that the start on such a program can be made only in the proper political atmosphere. The masses of people in backward countries do not understand either capitalism or communism or any other ism, but they do see American tourists and American businessmen, and they have before them their own rich and usually selfish classes. They are awakening rapidly to the fact that they do not have their own fair share of the world's goods. It is somewhat anomalous that rich Americans who have achieved their present level of prosperity through the capitalistic system are, by the examples of their behavior, helping to create a trend toward the forms of government which would do away with that system.

This brings us to one of the main points we wish to make concerning chemists. A young man spends seven or more years following high school to receive the specialized training required for the degree of Doctor of Philosophy. At the end of that time he is usually interested in science, and he is also fully aware of his worth to society. Hence he demands good scientific facilities, freedom to pursue his own ideas in research, and a good salary. Generally speaking, these individuals have attained little real social consciousness in the sense that their chief objective is to obtain a position which will gratify their interests in science rather than to be of real service. Those universities which provide research facilities have no difficulty in attracting competent young scientists because, even though they may not pay so much money as industry, they do provide that freedom for intellectual activity which many of them desire. Industry is a second choice for many of them in spite of higher salaries, because in industry personal egotism cannot be gratified by publication of results, and because the field of research must be restricted in general to something which is of interest to the company. Nevertheless, industry is able to attract a large number. Very few of these young scientists are willing to take positions, unless they are forced to do so, in small colleges with no research facilities, or to enter upon careers with the government with the idea of being of service either to the country or in a broader sense to humanity. A young scientist who takes a position where he cannot do research is making a real sacrifice, because the future will hold little for him in a strictly scientific sense.

The problem confronting the chemist who wishes to help the world to a better and more uniform standard of living, and hence to help in the cause of peace, is a very real one. The country needs more scientists who have those statesmanlike qualities which will carry them into positions of prominence in the governmental sphere. One can search very far to find cabinet members, members of Congress, and other high government officials who were trained in science. We like to believe that government is conscious of the power of science, but we must realize that while government is fully aware of the importance of science in waging war and does it lip service in certain other areas, it does not fully understand that scientific progress and international understanding are intermingled in a very real sense. Even when heads of scientific agencies are appointed, scientists are not always chosen.

We have in our government very many bright young men who desire to be of service, whose ideas, at least until they have become bogged down in the bureaucratic system, are of the highest. Very few of these young men in foreign affairs were trained in science.

We have made a hasty survey of chemistry and of chemists, and in so doing have touched upon a great many different and apparently unrelated topics. As a chemist I believe that chemistry has something to offer both in the cause of national defense and in the broader field of human relationships. As a nation we will go forward rapidly in applying science to the building of better gadgets both for war and for peace. Traditionally we have done little to provide the new ideas of science, and our achievements have been mainly in the fields of complicated gadgetry and of applied science. The first large generation of American scientists was trained immediately following World War I. Since that time we have been prolific in the production of scientists, but we have hardly had time to reach scientific maturity. We still do not have very many scientists trained in that abstract thinking which could lead to great ideas. In this area we have been followers and not leaders, and most of the men of that type in this country were born and trained abroad. We need more philosophical scientists and more positions for them to occupy. In chemistry in particular the number of theoreticians is extremely small. Many problems are begging for more really fundamental thought.

Thus, while we believe that chemistry has done and is doing great things, we also believe that a shift of emphasis toward basic science and toward the use of chemists and other scientists in the broad sphere of human relations is essential.

Scientists, and chemists are no exception, believe that they have a large fraction of the brains in the world. Many of them feel that they are true internationalists and that their training in rigorous thinking would fit them to solve the major problems of the world if they would only deign to bend their efforts in that direction, and if they were only allowed to function. This point of view is naïve and neglects some of the obvious factors which need to be considered.

A mathematician may make certain postulates, and by clear thinking arrives at certain inescapable conclusions. He has absolute control over the pertinent variables, and although we may admire his brilliance and the beauty of his work, we must admit that the difficulties he encounters are different from those in other fields of learning.

The physicist is able quite often to devise experiments

which show the effects of changes in one variable at a time. He is able to formulate many of his results in precise mathematical terms.

The chemist is in a more difficult position since the problems confronting him have more complexity and more difficultly separable variables, but he is still usually in a favored position compared to a biologist. The latter has made great progress, but his difficulties in determining the pertinent variables and in controlling them are often considerable.

Thus the mathematician, the physicist, the chemist, and the biologist have problems of varying difficulty, but all of them can, within certain limits, plan experiments with a view to ease of interpretation of results.

The social scientist is, in a general way, forced to approach his problems differently. Usually he cannot plan experiments, and he is forced to use data obtained without control of any variables. His conclusions must be drawn after he has tried to determine these variables and their effects. The impossibility of performing planned experiments makes it extraordinarily difficult to determine the effect of a single variable.

Perhaps we have oversimplified this comparison between various fields of learning, but I hope we have made one point clear. The attitude of mind of the natural scientist cannot be transferred bodily to social problems because the methods of attack are different. Nevertheless the natural scientist, properly oriented and open minded, may contribute to the solution of the complex problems in the field of human relations. Each group has much to learn from the other, and we need more cultured individuals who have some understanding of many fields. The scientist must, however, approach those fields about which he knows little with some humility.

Knowledge for its own sake is a worthwhile objective of science, but the ultimate goal of greater happiness for more people can be achieved only by the integration of many mutually tolerant disciplines.

CURT STERN

Genetics and the World Today

The scientists look at our world! Which world? The world of ideas? The world of human needs? The world of an authoritarian organization? Of a democracy?

Science appears different from different viewpoints, but none encompasses all its aspects. "The world" has always meant a multitude of phenomena, some apparently additive, others complementary, still others seemingly incompatible with one another. In the middle of the twentieth century we have become more conscious of this multiplicity of the world than perhaps ever before. If we look at science today we cannot afford to select some one of its aspects, but must view it from high ground and low. Genetics, the branch of science to which this discourse is devoted, may well serve as an illustration for several of the problems of science and the world today.

Let us begin with the adventures and delights of the spirit: Genetics as a *pure* science. Within the confines of the first half of our century, a body of knowledge in the field of heredity has been assembled, and a tool chest of concepts devised, which will always stand as a great accomplishment of human endeavor.

We all know the story of Mendel's successful thrust. He crossed a round-seeded to a wrinkled-seeded pea plant. All their offspring were round. He crossed the offspring among one another. Their progeny was part round, part wrinkled. He counted their numbers and found three round to one wrinkled. What of it?—one might be inclined to ask—and his contemporaries' reaction, or lack of reaction, is testimony to this shrugging of the shoulders. Yet out of this childishly simple couple of facts, the deep truth was lifted that the contributions of two parents to their offspring do not blend or merge into a unique hereditary newness but remain separable, to be recoverable unchanged in a later generation: clear-cut roundness and clear-cut wrinkledness.

Mendel noticed another fact. The round-seeded parent had yellow seed color, while the wrinkled parent plant had green seeds. Among the grandchildren four types appeared, with seeds round yellow and round green, wrinkled yellow and wrinkled green. Some of you will remember their proportions: 9:3:3:1. But that is a minor matter. The lever for further insight is the childishly simple fact that the parental traits round and yellow which came from one parent, and wrinkled and green which came from the other, had not always reappeared together in the combination in which they had been introduced into the cross, but had also appeared in the new combinations round green and wrinkled yellow. This fact reveals that each parent does not transmit a unified lump of hereditary matter, one whose joint consequences are in one case roundness and yellowness and in the other wrinkledness and greenness. Rather it shows that the hereditary matter of an individual is broken up not only into the two contributions of his parents, but that each contribution itself consists of separate and separable units. Thus the concept of the hereditary make-up as an assembly of many independent units was born. Just as the atomic theory had reduced the ephemeral multitude of chemical phenomena to the eternal properties of a limited number of atoms, so the theory of hereditary elements had reduced the multitude of specific individual appearances and disappearances of traits to the existence of a limited number of combining and disjoining elementary biological constituents.

This knowledge, not appreciated during Mendel's life, became our property at the beginning of the present century. What grew out of it are adventures of our time. Mendel knew that the separable hereditary contributions were contained in the plant's egg and pollen, or in the animal's egg and sperm, but he did not speculate about their nature or their specific location. A few further facts than those he knew have permitted us to place the hereditary elements into the small but microscopically clearly visible bodies of the cell's nucleus, the chromosomes. Still more facts, demonstrated first in sweet peas and fruit flies, have been the clues to the hereditary architecture of the chromosomes. While the chromosomes are contained in the nucleus in a haphazard fashion, like string beans in a bag, each chromosome itself is a well-arranged construction in which specific hereditary units occupy specific places. Thus we have accurate maps of the chromosomes of some organisms in which many of their thousands of different hereditary elements are assigned their locations: showing which one of the various chromosomes is the carrier of each specific element; picturing the linear order in which the elements are arranged within each chromosome; and giving the seriation of the elements and their relative distances within the chromosomal line-up.

What are these elements? Mendel himself gave them

symbolic letters, A and B and C, representing the characteristic traits which he was studying. There is no doubt that he did not think of A as being redness of flowers but rather as an agent transmitted through the parental germ cells, whose action, after growth and development of the fertilized egg cell into a mature plant, resulted in the presence of red pigment in its flowers. His early followers expressed this concept in the word "factor" (i.e., maker), the factor A leading to the production of redness. The famous term "gene" at first had no other connotation. It was to express the fact that the traits of the organism are *generated* by "separable, and thus so-to-say independent 'states,' 'factors,' 'units,' or 'elements' in the make-up of the germ cells. . . ."

The term "gene," however, did not permit itself to be confined in the lofty heights of hypothesis-free abstraction. When the gene was recognized as being associated with the chromosomes, an interpretation of its nature in terms of matter became an obvious need. We are still in the middle of this process of interpretation which takes place at the molecular level. Where the speculations of former times thought of elementary submicroscopic living units, themselves endowed with the mysterious properties which constitute life, we now regard the life of cells and of whole organisms as the resultant of chemical and physical processes which involve properties and reactions of molecular matter of the same kind which the chemist and physicist study in test tubes and with the spectroscope. It is a truism that this is not all. The mystery of life remains, but it is now seen in the integrated coupling of the molecular phenomena and the consequences of this overall molecular organization, not in the elementary processes themselves.

Thus our problem has doubled. While we want to know, justifiably, the gene's molecular structure, our task would not be completed with the attainment of this knowledge. If we could write down today an accurate structural formula of a gene molecule, the problems of genetics would not yet be solved. Indeed, we could imagine that the cellular chemist might have provided us with this structural knowledge before the science of genetics had been born. The molecular recognition would not have given us the gene, the generator of the organism's traits, but a molecule. The biological problems of genetics depend on molecular processes, but are larger than they.

As it is, the physico-chemical analysis of the chromosomes has not yet led to the gene. Beautifully refined micromethods made us recognize various proteins and nucleic acids, and even gave us specific arrangements of regions relatively rich or poor in these substances. It will be a long road, however, toward the recognition of the specificity of the thousands of linearly arranged entities which form the chromosomal gene string. We do not know whether the molecular analysis of the chromosomes will lead to the discovery of clearly separate specific entities, linearly arranged, and insulated from each other by less specific material. An alternative has been suggested, according to which a whole chromosome may be a unified kind of super-macro-molecule, and the genes only regions of specific chemical activity of parts of the whole. In this extreme form the alternative is not likely to be correct, since it is possible in the living cell to break a chromosome at many places and still retain the functioning of its parts. Interestingly enough, though, this functioning often is slightly altered from its normal action in the unbroken chromosome. There is some interrelation of neighboring genes, be it interpreted as interaction of separate molecular units or as interdependence of subunits of a larger molecular aggregation.

What started as a problem of passive transmission has become a problem of active functioning. Cell growth and development are now known to be founded on genic action. A simple experiment made seventy-five years ago, of cutting an ameba into two parts, one with the nucleus and the other without it, had shown the necessity of the nucleus for continued life and growth. Now the geneticist can show that the loss of a single gene may lead to damage and death. This new fact has provided a new tool to the experimenter. While it seemed hopeless to replace from the outside the activities of a whole nucleus, it became possible to discover the specific cause of trouble due to lack of a single gene, and to supplement the cell or the many-celled organism with a compound which it had lost the power to manufacture itself. Not that one can give such a cell a new gene, but at least we can supply a necessary product that is otherwise gene-dependent. And what is possible for one lost gene is possible for two and more. We do not yet know how much of the intrinsic groundwork of an organism we can do without as long as we take over its jobs. That one can go very far in this substitution is not a discovery of man's intellect. All through the evolution of life many forms have degenerated, losing their relative autonomy and becoming dependent parasites upon other creatures.

The role of the genes as key reagents in the biochemistry of the cell is being elucidated by the brilliant work of our contemporary biochemical geneticists. They have clarified many of the steps in which the genes take part in the synthesis of amino acids, vitamins, and other essential compounds. Different genes control or participate in successive steps in the so-called biosynthetic pathways. These studies have made genetics a central aspect of the science of cellular physiology which comprises nutrition and growth, energy transfer in respiration and fermentation, in synthesis and decomposition. And emerging from these phenomena, we are approaching from the aspect of genic action differentiation and development, maturing, aging and death.

As unbelievably remarkable as the functioning of an organism is under the guidance of its genes, we have not yet spoken of the still more remarkable fact that an organism can produce another organism, that it can reproduce. Once, in the century of rationalism and deism, some men believed that they could dispense with the solution of this biological enigma by laying it directly into the hands of the Creator. They thought that a man's reproductive cells, his sperm, were fully preformed little men, in turn containing inside their tiny bodies still smaller fully preformed human beings to constitute the grandchildren and so on in ever smaller proportion the encapsuled miniatures of all future generations. In terms of molecular biology this picture dissolves if it ever did have justification.

Reproduction of an organism, basically, has become reproduction of genes. Reproduction of a gene is an elementary process, the copying of an original molecular configuration within a cell. However dependent this copying process is on the preëxistence of the larger, living assembly of parts which constitutes a cell, genic reproduction becomes accessible to study by the physicist and chemist.

It is at the genic level, once more, that the problems of the evolution of living forms at present find their most fundamental treatment. If the genes of an animal or plant are responsible for its given nature, how is it possible that this nature changes in the course of biological history? If evolution implies deviation from conservative, accurate reproduction of an organism, the genetic interpretation demands deviation from conservative, accurate copying of the generators of living form. Genetics indeed has demonstrated that the so perfect copying process of the genic matter leaves room for some imperfections; that the stabilizing transmission of exact replicas of parental genes is accompanied by occasional upsetting lapses in which a new kind of gene originates and causes the appearance of hereditary newness. Some of these lapses are known to us in detail: rare cases where the copy of a gene remains with the original gene assembly instead of becoming part of the general duplicate; or rare cases where genic material gets lost from the genic assembly. Other lapses in exact copying may consist in a permanent change within the molecular make-up of a gene itself, a change which may have occurred before the gene began its reproduction, or may have been an error in copying itself.

The occurrence of such changes has been known for some time. They initiate *permanent* newness, and their discovery is possible primarily on account of the remarkable property that, once changed, further copies are reproductions of the new kind of gene, and do not return to the old pattern. We now do not need to wait for these mutations to occur spontaneously: we have learned to produce them artificially. The same century which brought us the artificial transmutation of the chemical elements has given us the artificial transmutation of the hereditary elements. X-rays and gamma rays from radium were the pioneer tools, and still are indispensable. Most recently, chemical agents such as mustard gas, formaldehyde, and urathene have been found which accomplish permanent genic alterations. The mechanism of induction of mutations is on the way to being clarified. Undoubtedly the solution of these problems will bring with it insight into the origin of the natural mutations which, uncontrollable at present, occur all the time in all organisms. It is clear already that these spontaneous mutations are elementary events dependent on accidental phenomena of nature: intrinsic instability of genic molecular compounds; unusual exposure of genes to reactive chemicals formed in the cell; and unavoidable events like natural radioactivity or cosmic irradiation.

These sources of mutagenic action are part of the forces of the universe. Life pays tribute to them, and harnesses them. The tribute consists in the sacrifice of many changed genes. Since the change is accidental in origin, much more often than not its result is detrimental. The mutated gene will not be suitable to take part in the established paths of gene action, and early death or weakness of the organism will result. This tribute is a tragic reality. While any one gene will mutate only in one out of many thousands of cells, the fact that each cell embodies many thousands of genes means that in each generation very many germ cells have acquired at least one mutated gene not present in their ancestry. In each generation millions of mutations are dumped into a species. Why then its apparent stability? The answer to this question has already been given. The early death or weakness of the individuals who carry the mutations removes the unstabilizers at the same rate as they are produced.

But not quite at the same rate. A few are harnessed to the yoke of survival. Are those a few good genes in contrast to the great bulk of bad genes? Molecules and genes are not good or bad by themselves. However, a few changes in the living assemblage of cellular molecules will happen to be compatible with continued good function, although the majority will not. The internally compatible new assemblages then will have a chance to be perpetuated. Whether they will coexist with the original ones, or whether they will replace them, will vary from case to case. Sometimes the new assemblage may thrive in a niche of the environment which had not been occupied before. The existence of thousands of different species in the same environment of a forest or a meadow is evidence of the existence of thousands of occupied ecological niches, and surely of many more yet unexplored. Sometimes the new assemblage will happen to survive in a changing environment where the original assemblage was doomed to extinction. The "deterioration of the environment," so-called by unadjustable men, is a continuous process; for the flexible, environmental change is a challenge to inventiveness.

Let us, however, not be deceived by this anthropomorphic phraseology. It fits our discussion in a figurative manner, but only in a most indirect way does the genetic model of evolution provide an organism with attributes of unadjustability or inventiveness. Changes at the molecular level all too often happen not to be adjusted to the demands of a higher level. Occasionally they do. The wonder is that for two billion years the genes have succeeded in escaping extinction. They survived, that is, left duplicates when they happened to clothe themselves in lambs and lions, amebae and bacteria, algae and oak trees, tapeworms and men. They perished when they were trapped within trilobites and dinosaurs, or within the tree ferns and giant horsetails of the carbon ages. The genetic view of evolution may be expressed in a variation of a well-known sentence: All living forms are only the genes' ways of making other genes.

Not that the genes could escape the process of evolution themselves. In order to create the superstructures of animals and plants within which to survive, the genes had to change themselves. Single, separately existing genes at the dawn of life had to become associated into harmoniously fitting groups, mutations of the genes had to make possible the development of the manifoldness of organisms. The one prerequisite invariant in the evolution of the gene had to be its property of reproducibility. What else remained constant we can only guess—perhaps protein or nucleic acid structure, taken in a general sense; perhaps enzymatic property. The evolution of the gene, as gene, is a field of inquiry barely touched.

This then is genetics as a way of interpreting living nature. This is Mendelism—Weismannism—Morganism—Mullerism — Goldschmidtism — Beadleism — Haldaneism — Fisherism—Wrightism. This is the framework which our times have built. It has been called "idealistic" by those to whom this term is one of condemnation, and with equal reproach has been termed "materialistic" by their opponents.

The new view of biology which genetics has provided may seem to many a denaturation of nature. Instead of birds with beautiful feathers and sweet songs, with artful nests and loving care of the young, we speak of the gene molecules which control pigmentation reactions and the deposition of cartilages in the singing box, determine the architecture and physiology of the brain, and the hormonal secretion of the pituitary gland. But the molecular interpretation of the individual development and functioning of an organism and of its historical evolution does not negate the existence of all the marvels of what we called the superstructure. The emergence of the latter is a miracle itself, and the exploration of life as it exists truly only above the molecular level remains as valid and tempting a task as ever before.

Our discussion of genetics up to this point has been subsumed under the heading "The adventures and delights of the spirit." The pursuit of such adventures and the tasting of their delights has had a respected place in civilization, not the least in our own Western tradition. "The world," wrote Sir Thomas Browne in the seventeenth century, "was made to be inhabited by beasts, but studied and contemplated by man: 'tis the debt of our reason we owe unto God, and the homage we pay for not being beasts. Without this, the world is still as though it had not been, or as it was before the sixth day, when as yet there was not a creature that could conceive or say there was a world. The wisdom of God receives small honour from those vulgar heads that rudely stare about and with a gross rusticity admire His works. Those truly magnify Him whose judicious enquiry into His acts and deliberate research into His creatures return the duty of a devout and learned admiration." Had genetics accomplished nothing else but its studies and contemplations, its position in the world today would be assured as one worthy of a mankind which not only wages war, invents nuclear explosives, and cannot learn fast enough to use its powers for good, but also creates and supports seats of learning and thought, preserves and defends some of the beauties of the world, and, in spite of all, succeeds with innumerable unselfish acts of duty and love.

Yet man liveth not of adventures and delights alone. Genetics has had the good fortune to fulfill not only the social function of exploring its realm of the universe, but to contribute to our material well-being and provide foundations for social measures. Let us now consider some of these material products which we may call "the produce of the mind."

The most famous of the applications of genetics to the practical world of today is the technique of raising hybrid corn. It had always been the goal of plant and animal breeders to produce strains which have two properties:

one, to give high yields; and two, to be pure. This latter trait, purity, would assure the homogeneity of the strain, that is, the likeness of all individuals grown at the same time, *and* the constancy of the strain, that is, the likeness of parents and offspring. The likeness of the individuals of the same generation would assure equal potential performance or yield from all of them, the likeness of successive generations would enable the farmer to continue his stock through the years. The radical innovation of hybrid corn retained the goal of homogeneity of one generation and increased immensely its yield. For this increase it had to pay a price, the renunciation of likeness of successive generations. The technique of raising hybrid corn arose from a theoretically and experimentally inspired contemplation of the composition of a field of corn as it used to be grown all over the world. It became apparent that the varieties of corn then on the market were far from being homogeneous. When genetic procedures were used to obtain pure lines they were successful. Many different pure lines were produced, each one homogeneous within and between generations—but none was of practical value. Always growth was stunted, and fertility low. This appeared strange enough, but stranger still, and happier, was the outcome of crosses between the poor, pure lines. The hybrids not only showed better vigor than their parents, but they surpassed the desirable qualities of the individual members of the old field of corn.

That the desirable crossbred plants were undesirable as producers of seed for future generations lay in the nature of the genetic make-up of the hybrids. The shuffling in reproduction of the parental genes would result in renewed heterogeneity of the next generation, and in loss of the specific hybrid vigor. Thus the discovery which had yielded such desirable produce requires each year anew the effort of hybridization of the, by themselves, undesirable pure strains.

The theoretical problems of hybrid vigor are not yet solved. On the contrary, the cause of this phenomenon is a matter of extensive research at present. The practical fact is that after a slow start with the first commercial seed field of hybrid corn grown in 1921, and a gradual rise to 80,000 acres in 1932, now the greater part of the nearly 90,000,000 acres of field corn and sweet corn in the United States are planted with hybrid corn. It has been estimated that perhaps a billion bushels a year are now harvested in excess of the best national yields we enjoyed before we had hybrid corn.

Genetics is credited for this practical result because of the historical fact that a pure geneticist, G. H. Shull, working at the Station for Experimental Evolution on Long Island—not at an agricultural experiment station—was the father of hybrid corn. Admittedly it was no absolute internal necessity that the introduction of hybrid corn into practice had to wait for the concepts of genetics. Indeed, the use of crossbred field corn had been proposed many years earlier, while mules, the vigorous but sterile hybrids of the horse and ass, had long had a recognized place in animal production. Yet it seems that the modern intellectual penetration into the phenomena of heredity was required to rediscover, refine, and assure the acceptance of the unconventional procedure.

The success of applied genetics with hybrid corn is followed closely by the success of the breeder of disease-resistant wheat plants. Each year the loss of yield due to widespread infection of wheat by parasitic fungi used to be a drain on the national economy and, even worse, on the source of livelihood of many farmers. Then genetically trained plant pathologists discovered that susceptibility to infection was not a ubiquitous fate. Specific genetic constitutions were present in many plants which endowed them with resistance to the invaders. Varieties were bred which not only possessed those desirable properties for which they had been selected by plant breeders, but which combine these properties with resistance to infection. In this way for many years whole regions suffered little loss in the yield of this important crop. Genetics, furthermore, served to explain and to prepare the farmer for the breakdown in protection which again and again brought catastrophe to the wheat grower. A gene that protects a wheat variety is not all-powerful. The fungus itself is heterogeneous, and different genetic types of the parasite are repulsed by different genetic types of the potential host. Invasion by a new strain of the fungus into a territory where no resistance against this specific strain existed, seems to undo past successful efforts. Even worse, mutations in some of the myriads of spores produced by even a few fungi may lead to the origin of new virulent varieties. Thus the plant pathologist and geneticist has a never-ending task, but he can anticipate a future unavoidable genetic breakdown of any specific resistance mechanism by preparing new varieties of wheat, resistant to new laboratory-cultivated varieties of the fungus.

This recognition of parasite mutability strikes still nearer home. Bacteria have been found which are resistant to sulfa drugs or streptomycin, though their ancestors could have been combated successfully with these drugs. Even streptomycin-dependent bacteria occur which thrive only on what means death to their brethren. The origin of these new types is due to mutations in the genetic composition of the bacteria. One of the newest branches of genetics, microbial genetics, is now elucidating these phenomena and contributes to overcoming the impediments to drug therapy.

It happened that at about the time when the artificial production of mutations was first accomplished by means of radiations, the action of the penicillium mold in inhibiting bacterial growth was discovered. Years later, during the last war, when the large-scale production of penicillin became an urgent task, the geneticist's method was applied to the production of strains of the mold with greatly increased output. The strain now used most widely by commercial firms originated from irradiating spores from relatively low-yielding strains and selection of a spore in which a mutation to high yield had been induced. Similar methods of causing mutations responsible for desirable new qualities in agricultural plants have been employed, particularly in Sweden. Success has been slow in coming, for reasons which might have been foreseen. In evolution, establishment of new genetic types results not directly from the occurrence of a new, mutated gene variety within the genic assemblage of an individual. Rather, it requires a process of selecting some of the varied genic assemblages of the species, namely, those which happen to complement in a harmonious way the newness of the mutation. Similarly, in artificial breeding the production of a useful mutant requires a selection of suitable genetic background. This adjustment between the new and the suitable elements of the old takes time. By now, improved strains with superior stiffness of the straw in barley, and an improved strain with superior seed and oil yield in white mustard, have been created on the basis of X-ray-induced mutations.

They were pure geneticists who discovered the peculiar mode of inheritance of some special traits which are called "sex-linked." In suitable crosses with such traits all daughters resemble their fathers and all sons their mothers. It was another pure geneticist who proposed to use this criss-cross inheritance for the sexing of chicks. Male and female chicks are hard to distinguish until they are several weeks old. But when you cross barred Plymouth Rock hens and not-barred Rhode Island Red roosters, the female chicks are not barred and the male chicks are barred. This trick is now being used each year for the commercial production of several million chicks in the United States alone. Obviously, only females can be used for the all-important egg production, and it is more economical to destroy most of the newly hatched now easily recognized males than to feed them until their normal development betrays their relative uselessness.

One more example from agriculture. Cattle and horse breeders as well as breeders of other animals have always been troubled by the birth of abnormal types of young, often doomed to early death. When genetics arose it showed the frequent occurrence of lethal genes in mice and in fruit flies, genes that lead to abnormal, destructive development of their bearers. Soon it was recognized that many of the stillbirths which had troubled the animal breeder were due to specific lethal genes, often brought into a herd by some famous sire who carried the lethal in a harmless combination with a normal gene. Among his numerous descendants many would be carriers again, and breeding the carriers with one another would result in 25 per cent of the pregnancies ending in disaster. This insight into the stock breeders' troubles carried with it means of avoiding them. It was easy, on the basis of Mendel's first law, to devise trial matings through which bulls or stallions could be recognized as carriers of dangerous lethals and, if found genetically unsound, be excluded as sires of large numbers of offspring. The savings accomplished in this way are very considerable. Even in so inexpensive an animal as the chicken the recent discovery of a particular lethal gene has bared losses, now avoidable, which in a single hatchery amounted to more than $180,000 over the last ten years.

The most significant applications of genetics concern ourselves. Men are born genetically unequal. This is a fact of nature, and quite independent of the conclusions which may result from its political and sociological interpretations. Man is therefore a subject of genetic investigation, and much has been learned about him in our time. The produce of this knowledge extends both to the individual and to the community. How much personal worry is relieved when the human geneticist can advise a healthy questioner that his or her chances of having normal children are as good as anyone else's, in spite of the fact that perhaps the father, his mother, and several brothers and sisters have been afflicted with some serious abnormality. How much further suffering has been avoided when the genetic counselor had to predict the high probability of a sad affliction reappearing in a family, should a new pregnancy be attempted. Human heredity clinics fulfill a great need, and are still all too few.

Yet more important than such advice from case to case are the applications to policy. The urgent warnings of geneticists against careless use of X-rays and other ~~ionizing~~ radiations may well have prevented the production of thousands of human mutations bad in effect, as most of them are. In today's world of atomic energy use, the need of shielding workers from radiation has taken on greater significance than ever.

The case against careless irradiation transcends the interest in the immediate offspring of exposed persons. Many induced mutant genes will not show their effect in the first generation but at any time, far into the future. This aspect widens the responsibility of a world today beyond its usual care. The well- or ill-being of our distant descendants is, to some extent, in our power. This power is not restricted to the yet minor aspect of radiation genetics. If men are unequal genetically, then our actions and inactions are bound to influence the genetic composition of the future human populations.

The applications of genetic knowledge to this great problem are largely negative at present in that they serve to expose misconceptions. It is important, for instance, to debunk authoritatively distributed pamphlets which endeavor to encourage large families by the misstatement: "Heredity favors the third, fourth, fifth and subsequent children rather than the first two, who are apt to inherit some of the commonest physical and mental defects!" It is important to rectify the opinion that political and moral equalitarianism has any bearing on the biological facts of man's genetic diversity. It is equally important to stress the great adaptability and plasticity of man, particularly in mental attributes. Much of the genetic diversity may play an insignificant role in the actually observed range of normal and even abnormal human behavior and accomplishments. The results of genetics are fully compatible with the recognition that many men with equal accomplishments may greatly differ genetically, and many other men with different accomplishment be much alike. Genetics gives little basis for beliefs in basic distinctions among classes and races.

Those who in their hearts applaud the last statement in general often are loath to accept the relative term "little" which was used. "No basis," they feel, would be a more fortunate expression. The observing scientist, however, has no power over what is fortunate or unfortunate. If he comes to the conclusion that there is no difference he will say so, but if his finding is "some difference, even if slight" he cannot falsify the record as he reads it.

One of the most comforting results of the geneticist's thoughts has been the recognition that the problems of the genetic future of mankind are not as urgent as they seemed a generation ago. Mankind will not degenerate overnight if nothing is done to change its reproductive patterns, nor would it have blossomed out suddenly if the old-time eugenicists had had their will. We now know that the immense genetic pool of mankind can be changed only very slowly, for better or worse. Here then is one problem at least about which mankind can take its time.

A problem it is, nevertheless, and eugenics, in a sobered mood, is still a demanding goal. While it is true that the methods of the animal husbandman should not be applied to human beings, the discovery of man's genetic diversity cannot escape the treatment which men have successfully applied to so many other facts, namely, use in intelligent planning. It is strange that resistance to this proposition seems strongest among some of the most audacious social designers.

Genetics in the world today can point to spiritual flowers as well as to material fruits. It can justify its existence on either ground. Nor does it need to be ashamed of either. The student of universal phenomena may well rejoice when his search leads to practical benefits, and the biological engineer may well appreciate the more detached task of his theoretical colleague. We cannot afford to measure all human endeavor by its practical benefits. Where would painting and poetry be, where astronomy and archeology, where games and hiking and sports? If we do not permit these activities for pure joy's sake alone, we shall end up with the slogan *Kraft durch Freude* (Strength through joy) which historically turned strength into disaster.

The enjoyment of the arts and the sciences, of nature and life, is the individual's privilege, but it can be thwarted or encouraged by a social control. The munificence of princes made possible the painting and sculpture of the Renaissance, that of the wealthy men in our country the foundation of research institutes and of well-endowed centers of learning. Now that learned activities depend more and more on society at large, on its delegates to the Congress and the state legislatures, an enlightened understanding of the spiritual significance of thought and search must be kept awake and be strengthened over the widest possible ranges of our citizenship. Should that not be possible in a country in which almost one-fifth of the young people between eighteen and twenty-four years of age are enrolled in school or college?

Genetics has a particular reason to sound a warning. Its recent fate in Russia was intimately bound up with a disdain for the humanistic aspects of science. It is unavoidable, in any society, that the support of humanistic activities is subject to general conditions. The demands of economics, of a mobilization, of changing social needs, will create forces which divert at one time more, at another time less manpower and funds to a given function. There would not have been necessarily dangerous implications had the Soviet Government decided to curtail to some extent the expenditure of its intellectual and financial resources which had been devoted to genetics. It would probably have meant bad judgment about the long-term advantages to be gained from asking experts in basic research to turn to more immediate tasks. What is so distressing was the contempt for scholarship and abstract—yet real—

thought which motivated, not the curtailment, but the suppression of genetics. In his "victory" address of 1948, Lysenko made this clear once more when he emphasized ". . . what led me to study profoundly theoretical problems . . . was never mere curiosity and a fondness for abstract theorizing," and when he held up to the ridicule of his audience a basic work of one of the most brilliant of his countrymen.

We ourselves are by no means immune to influences which regard the activities of the spirit as dispensable luxuries. Let us remember the sober words which Washington wrote in the war-torn America of 1780: "The Arts and Sciences essential to the prosperity of the State, and to the ornament and happiness of human life, have a primary claim to the encouragement of every lover of his country and of mankind."

ALAN GREGG

The True Outwardness of Medicine

As a fanciful invention, let me ask each of you to imagine that he will be host next week at a dinner party for twelve—a dinner party satisfying the highest standard of hospitality, namely, that every guest be seated between persons certain to delight him and sure to kindle his affection. Those of you who are experienced hosts or hostesses will have already perceived how fanciful are these mere stipulations for a dinner party, but if you will strain your imaginations a bit further and agree to but one more condition, then an uncommonly entertaining affair is assured. This condition of success lies in the choice of your guests. On your right you will place your father when he was just your present age, on his right his father at that same age, and so on all around the table; a direct line of son to father back for eleven generations. Imagine it! Don't upbraid me with the dull protest that it cannot be. The idea is real and, like the constructs of the mathematicians, or the subjunctive mood of many a language, concepts that are mere suppositions provide ofttimes our only way of dealing with reality. You ladies may prefer to invite your mother, your mother's mother, your great grandmother, and so on, following the matrilineal side—all, however, at your present age.

Despite such visionary hospitality, I can lay claim to being practical in one point: that of suggesting that only the fathers' fathers or the mothers' mothers be invited, since in the eleventh generation back, each of us has a total of 1,024 grandfathers and 1,024 grandmothers to claim—proudly, I hope—their equal biological shares in our existence. And if all the intervening forebears, male and female, back to, but not including, the eleventh generation were to be invited, you would have still another 2,046 for the cook to reckon with. A dinner party of twelve with only one to represent each generation would therefore be an extremely exclusive affair—the ideal of many a hostess, and economical, too—the ideal of many another. Allowing an age span of thirty-five years from generation to generation, the man on your left would have been born about 1511 and therefore have attained your present age somewhere between, say, 1536 and 1571. Except for yourself and the man on your left, each guest would be seated delightedly between his father and his son—and the uniformity of their ages would help the conversation. Only between the man on your left and yourself would conversation be perhaps difficult. Indeed, it might be impossible. In my case the guest on my left, a lowland Scot named MacGregor, would be speaking with a well-nigh incomprehensible Scot's burr, if indeed he were not confined to Gaelic. With presumable Scotch dubiety he would be inclined to distrust such items on the table as potatoes and ice cream and coffee—completely unknown in his day in Scotland. And he would be totally alarmed at cigars and cigarettes, and the sight of his own descendants evidently on fire within and certainly smoking at their very nostrils.

Spellbound as you might be at the spectacle of your

guests' clothes and their table manners, and awed as they might be at such table finery as royalty itself had not had in their day, 'twould be a cheerful and mellow party in point of conversation. What comparing of notes, each situated between his father and his son, the father for once equally young, the son at last equally mature! And every guest but the man on your left placed between persons, as I promised you at the outset, certain to delight him and sure to kindle his affection anew. And even this biologically young but historically antique grandfather at your left, if I understand grandfathers, would be speechless with pride and wonderment at the array he had, as men like to think, produced, and the marvels that God hath wrought.

Tonight, in an evening deliberately devoted to the memory of Benjamin Franklin, and adventitiously recalling George Washington,[1] I could easily spend the time in historical reveries and imaginings as to the conversations that might take place. What variant opinions! On original sin and infant damnation, on the divine right of kings and the French Revolution, on farming practices or education, on the future of the Colonies or the works of the new playwright Shakespeare, or the discoveries of William Harvey. But I am no historian, and besides my hope is rather to make you feel the general circumstances of medicine, its sweep and movement, its interest and its fiery vitality. There is not time to reconstruct many of the possible conversations between fathers and sons all the way around our imaginary dinner table, vivid and revealing as they would be. Let us then change our mood from fantasy to realism. Let us discuss the changes in the attitudes toward illness and health that onlookers after years of study would distill from the conversations at such a meeting of your ancestors.

As the word "comprehension" suggests, we can under-

1 This address was delivered on Washington's birthday.

stand anything only when we know what surrounds it as well as what it contains, its circumstances as well as its content. To seize the true outwardness of an event leads to a better grasp of its true inwardness. So I take as my title "The True Outwardness of Medicine," for I want to talk about what surrounds both the use and the advancement of medicine—in short, of man's attitude toward disease, his interpretation and understanding of suffering and pain, decrepitude and death, and of the judgments of value we acquire and apply in order to get well or stay well. True, changes in these attitudes toward disease and health have often come out of empirical discoveries and scientific research. But almost as often what man expects from medicine has come from other quarters, from the religion, the philosophy, the *Zeitgeist* of his day. In such matters this audience may be presumed to have a greater and more natural concern than in hearing a long recital of names, dates, and technicalities in the development of medical science.

The *why* of medical practice and research deserves quite as much attention as the *how*. There is a wise aphorism that runs, "The man who knows how will always find work; the man who knows why will be his employer." If you are to employ or use doctors wisely, it behooves you to know why. I want to show you some of the whys of medical practice, medical education and research, for it is upon the estimation and understanding by society as a whole that medicine depends. What, for example, is our attitude toward the sick?

In the long, slow growth of its art and science, medicine owes an incalculable debt to the part of the Christian ethic that enjoined upon its adherents charity and assistance to those who suffer. Resulting from this injunction, God's hostels, later known as hospitals, were created and maintained. Instead of leaving the sick as isolated, deserted pariahs, they accomplished something new—they brought the sick together. Thus it became possible and even easy to compare illness in one person with illness in another. This in turn laid the way for recognizing the same disease in different patients, and then knowing different diseases apart, a cardinal procedure now known as diagnosis. In short, a religious belief enormously affected the study of disease and the progress of medicine.

Along with this tremendous help from the Christian religion there were some considerable handicaps, also of religious origin. Pain and disease, as was the case in other religions, were interpreted at times as the punishment of a vengeful God or the will of an inscrutable one. For example, the year my father was born, 1847, Sir James Y. Simpson incurred the determined wrath of the Scottish church for his use of chloroform in childbirth. Did not the Scriptures indicate that woman is ordained of God to bring forth in travail and in sorrow? Did a mere doctor dare to interpose a new-found chemical in contravention of woman's natural lot? For an example of the punitive interpretation of disease, I need consult only my own memory. When I was an intern, admission of a patient with recently acquired syphilis was refused in my august teaching hospital, though other acute and dangerous infections were freely admitted, and cases of tertiary syphilis dotted, if not indeed pocked, the wards—a refusal that suggested more of puritanical vindictiveness than of preventive medicine. The patient's occupation, to which he presumably returned from the closed doors of the hospital, was that of preparing the salads in Boston's then most fashionable hotel.

At no great distance from regarding disease as the Divine punishment of sin "even unto the third and fourth generation" stand the myriad thickets and morasses of superstition from which medical science has slowly extricated us. So powerful are the terrors of the beginning and the end of life, and often of its intervening accidents and illnesses, and so comforting at such times are credos and traditional, or at least approved, forms of procedure, especially when supported by superstition's flimsy logic, that we can rightly wonder how man ever did make his escape from mere superstition to a more rational handling of life's crises. And lest you think I imply that superstition is quite dead, let me ask you this question: Haven't you ever heard anyone say, "Fortunately, I have enjoyed good health all my life"? Chance, of course, still plays some role in the incidence of disease and death, but I venture to say that without sound medical knowledge and its application over the past century to your grandparents, your parents, and yourself, a quarter of this audience would either never have been born or would not be alive or stalwart enough to attend a Franklin lecture. And I happen to know that if my grandfather had not taught himself how to put on a tourniquet, you would be listening to a different speaker. To think that your own benefits from the general improvement in the incidence and fatality of disease is a matter of luck, is hardly better than pure superstition.

To most of the guests at your ancestral dinner party the idea of the control and prevention of disease would have all the flavors of a dream, except that control or prevention of disease can be proved to be a reality today. Even if this is not yet true for the common cold, poliomyelitis, and some types of cancer and other current enemies not yet understood, it is true for typhoid, smallpox, surgical sepsis, tetanus, childbed fever, malaria, dysentery, yellow fever, hookworm, typhus, diphtheria, tuberculosis, scurvy, pellagra, and other diseases of malnutrition, and the long list of conditions such as appendicitis whose final effects can now be averted by safe surgical intervention. All these controllable diseases have existed in this city. Indeed, without today's dependable knowledge applied to control and prevention of disease, no city the size of Philadelphia could long hold its citizens, panic stricken sooner or later as they would be by the epidemics that would flourish in such a mass of unprotected victims.

Another extraordinary change of attitude toward disease and its care begins to distinguish the present from the past. The services of a physician used to be considered the privilege of the rich or a priceless boon given to the poor in the name of Christian charity. In the last hundred years they were even offered as a gift to the "heathen" by confident medical missionaries in the name of the Lord. Today the dean of the Harvard Medical School writes, "Freedom of opportunity for good health is a basic human right. That this opportunity for good health should be provided for all is an insistent and growing demand. It is a major social and political fact today."[2] Demonstrate the effectiveness of medical care and preventive measures in wartime to some four million young Americans in World War I and to fourteen million more in World War II, maintain the Veterans Administration as a follow-up, and now draft physicians to look after three and a half million young conscripts in time of peace, and you will have passed the stage of theorizing about the value the voters will attach to having medical care. Medical care passes from being a privilege of those who can pay to being considered a civic right of those who will vote that it be paid for. As in the case of clean water or education, what starts as private or proprietary initiative passes through stages of privilege and contracted service to being regarded as a civic right

2 *Union Worthies,* Series No. 5, Union College, 1950.

and a function of government. The private practice of education was not terminated by the socialized education we call our public school system. Nor will the private practice of medicine be anything but improved over the years by the extension of medical care to those who cannot pay except through taxation.

When your great grandfather sought to take his brother to hospital in the agonies of "acute indigestion" (i.e., appendicitis as we know it now), the family, in horror at this last step before the charnel house, prevented it, saying it was better to let Death come to our house, if he must, than to court Death in his house. And they had some reason. The hospitals were unhealthy places even for women in labor, as Semmelweiss showed. Until Pasteur and Lister, during your father's lifetime or your grandfather's, showed how wounds and surgical incisions became infected, hospitals were places where fatal infections could be not only treated but acquired. Indeed, there is a surgeon still alive in Massachusetts who recalls that he once offended his fiancée because he pinned a rose in his lapel with a pin from one of his surgical cases, and the pin, still reeking with laudable but noisome pus to which the young surgeon was accustomed, o'ercame the fragrance of the rose. The fear of the hospital has almost gone—another change of attitude that argues well for future progress in healing and our mastery of suffering.

But more significant than these vivid evidences of progress in the beauty and grace of living is a change in attitude described as follows by Freya Stark in her book *The Southern Gates of Arabia.*[3]

> In the West, spasmodically and with uncertain hands, we try now to eliminate the causes of sorrow: but it is only recently and since the decline of formal religion. The East still holds

[3] New York: E. P. Dutton & Co., Inc., 1936, p. 143.

religion in its established forms: and encourages philanthropy which deals with effects and not with causes. For as soon as you investigate and try to alter the origins of things you are no longer a philanthropist but a revolutionary, and your disinterested movements are liable to make whole edifices crumble: and mankind is asked from successive pulpits to leave the fundamental things alone.

Medical research in its constant quest for the origin of things is revolutionary too, though most of its triumphs have brought such comfort, relief, and reassurance that its revolutions are agreeable, honorable, and almost soporific. But let medical research approach such subjects as sex, psychiatry, socialized medicine, heredity, or the control of reproduction, and "mankind is asked from successive pulpits to leave the fundamental things alone." Indeed, if communion between the generations were possible, as at your imaginary dinner party, I imagine that the one likely source of discord might lie in the horror and aversion of each older generation on learning what their damnable descendants were inquiring into with Promethean insolence.

The discovery of research as a method of asking revealing questions of Nature is really the discovery of how to make ever more discoveries. For this reason the tempo of the resultant gains in knowledge resembles a chain reaction. Ten or eleven generations ago a conversation between father and son could have covered in ten minutes every important and lasting medical advance occurring in the lives of either father or son—and that is on the dubious assumption that either was even aware of any such gains in knowledge. In listing the great advances merely within my lifetime and that of my father, I actually go back to 1847 when the term "anesthetic," a neologism coined by Oliver Wendell Holmes, was scarcely two years old. The sequence of fundamental discoveries in medicine in the last hundrd years waxes in importance, and the number increases in each successive decade. Obviously, such formidable discoveries have not been confined to medicine: witness the telephone, the automobile, the airplane, the radio, radar, television, atomic energy—the pace quickens. With what general result?

The late Elton Mayo, once a teacher in the University of Pennsylvania, stated one highly significant result: our civilization is no longer traditional, it is adaptive. A heavy rain that follows seeding the clouds with dry ice or silver iodide challenges the traditional view that nobody is responsible for the weather: lawsuits brought by those whose crops were ruined against those whose crops were to be benefited call for adaptation in point of one of our oldest traditions, namely, that a rainstorm is an act of God. That word "adaptive" has many implications. Among other things it means that our way of life is changing so rapidly that the experience, the wisdom, or the value-judgments of the older generation cannot be passed on, as before, for the guidance of their children. Not even when such wisdom is asked for! The die-hard convictions of an earlier generation have, indeed, so often underestimated the potentialities of a new invention or discovery, that age itself becomes discounted, more because of its refusal to see than because its eyes have become blurred and rheumy. Even today we cannot recognize change quickly enough to adapt ourselves to it. Do you wish an example from the recent progress of medicine?

Lawrence Henderson described in an interesting way the progress of medicine. He said he thought that somewhere about the years 1910 to 1912 in this country it became probably true to say that a random patient with a random disease consulting a doctor chosen at random stood better than a 50-50 chance of profiting from the encounter. I hesitate to think that a random patient in 1910 can be assumed to have had money enough to call a doctor at random. But there are stories of that day to suggest that doctors' bills were not always paid; so perhaps he called a doctor anyhow. Considering the progress of medical discovery since the turning point of about 1910, we may take it as axiomatic that the doctor and the surgeon today, if they are well trained, have now far more to offer, and with certainty to deliver, if they are called. And note that Henderson confined himself to the advances in what a doctor can do when he is called. The prospect of what medicine can do is far more remarkable when the doctor today is permitted by law occasionally to take the initiative, as he has in the prevention of some diseases. In prevention his services have been as astonishing as they have been underpaid and underrated.

Do you realize that for six Americans in service who died per year from disease in the last war, one hundred and sixty died in World War I, and two hundred and fifty in the Spanish-American War, and six hundred and fifty on the Northern side in the Civil War? And lest some cynic sneer that it would only be tax-supported medicine and war conditions that could show such remarkable improvement, let me cite progress in the city of Chicago over a far shorter and more recent period. Twenty years ago Chicago had 6,012 cases of diphtheria and 513 deaths; last year five cases and no deaths. The full list of such triumphs of scientific medicine is long. The very dependability of medical science leads us to take it all for granted. But is such easy acceptance a wise attitude even if it is characteristic of our omniscient times?

The casual apathy of the prosperous disturbs me. Do you think good health is a matter of luck? It is no more a matter of luck than good food, good clothes, and good housing.

There are still, it is true, many kinds of disease of which we are ignorant and therefore almost powerless to prevent or control. I'll grant you that whether you are to die from a disease we don't yet understand has something of luck about it, though even there the dice would be loaded in your favor if you would support medical education and research "for dear life," as one might truly say. But in the light of what medical science has done already and stands ready still to do, I beg you to stop thinking that good health is a matter of good luck. God gives us hunger to remind us not to starve to death—and we budget for food. God gives us the shivers to remind us to build houses and wear clothes so as not to freeze to death—and we call these basic living expenses. God gives us brains with which to combat death from disease—and we haggle over the price of adequate medical care and let our medical schools languish and retrench for a paltry deficit of ten million dollars in 1948 in a country that spent more than four billion one hundred million dollars ($4,100,000,000) on tobacco and smoking supplies during that same year of grace. What a fantastic indifference to real values!

Do you wonder that I have put my emphasis on the true outwardness of medicine, that is to say, the prevalent attitudes and values you have about the utilization of medical science? There is an extraordinary lag of the laity in these matters. Our forebears, if they were alive today, would simply not understand it. They would not understand why, if we know how to prevent disease and educate doctors, we do both in such picayune and reluctant fashion. Imagine the expression on the usually benign countenance of Benjamin Franklin, the publisher, if he were told that the Associated Press devoted only 1.7 per cent of its wire service during a recent period to the combined subjects of health and science! His electricity so churlishly engaged!

Nor does the fault lie only with the laity. I do not understand why the deans of our medical schools have not served joint notice on the American public that without further support they cannot be held responsible for continuing the recent quality of medical education. In hospital language, American medical education should be put on the D. L.—the Dangerous List—as a warning by those who are responsible to those who are concerned. Niggardly salaries, crowded classes, and inadequate facilities will never attract and are at present discouraging younger teachers. In our American medical schools there are 240 existing teaching positions vacant—vacant for lack of money now, but soon for lack of trained young teachers. And if, as is frequently reported, the American Medical Association used its funds or influence to prevent federal tax support from going to our medical schools last autumn, I do not understand that either, in any terms that have to do with the general welfare in times of national emergency.

I sometimes wonder whether our medical profession has not been stamped and molded more than we realize by having, in our care of patients, to start nearly always with a handicap—the handicap of being called not in time, but a day, or a week, or a year, after the disease has already established itself. Our work so often demands that we take things as they are, that we salvage what we can, that we refrain from recriminations that won't mend the situation. The result is that we take what is at hand and do not ask for what is ideal lest it prove impossible. Unfortunately, we learn to shrug our overburdened shoulders. Add the fact that the patient wants to have confidence in us: the result is a doctor who soon loses the capacity to plan for or even imagine the perfect end-result and, like the western organist, simply does the best he can with, meanwhile, what looks like an extreme sensitiveness to criticism from any quarter. It might be profitable to study the broad effects of the conditions of work of each profession on the men who enter it. I recall the shrewd but kindly advice given to us as senior medical students in 1916 by George Gordon, the minister of the Old South Church in Boston. We asked him to talk to us about doctors as seen by a minister of the Gospel. This was about what he said:

I have often observed how remarkably a man's professional associations eventually affect his manner of thought and feeling. Now I have served on many civic committees where I have watched the behavior of professional men and it would be my considered judgment that if you are ever on a civic committee of one kind or another you'd do well to respect the professional experience of the members of the committee. You'd do well to divide the work into different parts according to the professions represented.

For the work of examining the status quo, no matter how bad it is, and making a report on it, you'd do well to use the doctors on the committee. They're honest men trained to observe and in touch with all parts of the community, accustomed to knowing the worst and doing the best they can. But doctors have one great defect. They've been so surrounded by frightened patients, adoring families, and obsequious nurses that they're not accustomed to being criticized or having their recommendations challenged by any living man or woman. So take the doctor's report arbitrarily out of their hands and turn the duties of discussing what's to be done over to a committee of lawyers.

Now the lawyers have one great quality; they have the hides of rhinoceri. They are quite capable of criticizing each other's suggestions mercilessly all the afternoon and still dining amicably together in the evening—which is more than the doctors can do. But the defect of the lawyers is that they would be willing to go on that way forever. So, when the fifth plan suggested (and amended twice) begins to look suspiciously like the first suggestion that was made, it's time to take it out of the hands of the lawyers and hand it over to the businessmen on the committee. Now the businessman is congenitally incapable of examining anything and he will recklessly override the rights and feelings of others, but he has one grand quality—when given a clear idea of what he is to do, he will seize upon it like a dog with a bone and carry it through.

As the different occupations and professions in our modern world become more specialized and separate, can we say that communications have become such as to counteract their separation from each other and the rest of us? Our newspapers, for example. Dr. Brock Chisholm, the head of the World Health Organization, told me last September that when at a press conference he gives an account of agreements arrived at among the different nations, of difficulties finally ironed out and problems solved, the reporters tell him that they want *news.* By news, they mean news of quarrels, fights, resignations, plots, scandals. Hasn't he any *news?* To this implied criticism of the public prints I know the professional newspaperman's reply, "We only give the people what they want." Though it reminds me of the meretricious self-defense of the oldest profession, both are right in placing the chief blame on the consumer. You make what you buy, even for your leisure time. Indeed, nothing shows the culture of a person so unmistakably as his use of leisure time. The Greek word for leisure was σχολή from which was derived the Latin *escola* and our words "school" and "scholar"—surely an eloquent testimony to what the ancient Greeks were at heart and how they chose to spend their leisure time. Will our phrase "time off" ever be borrowed by some future language to describe activities of the mind and spirit? I doubt it. Our passion for cheap emotional orgies, for *Schadenfreude,* for news of someone else's disgrace, frustration, failure, shame, sorrow, or murder, leaves us often but one page in the newspaper devoted exclusively to solid, honest accomplishment, and that, ironically enough, is the obituary page.

We buy such newspapers for our leisure time, especially on the Lord's Day.

In mentioning the financial deficits of our medical schools, I do not wish to imply that all they need to do is get back to their prewar efficiency. There are new fields they could enter. There is plenty still to do. If there were not, I'd just be sorry for the younger generation. For example, medical science could be applied to the detection of crime and the protection of those innocent of crime but tragically accused of it. While departments of Legal Medicine in medical schools in Boston, Baltimore, and New Orleans struggle and strain for as much as $30,000 a year to serve the desperate cause of human justice, the American public buys over a million murder mysteries a year at probably two dollars and a half apiece and some tens of millions of reprints at twenty-five or thirty cents each. Thus we spend probably ten million dollars for vicarious thrills over imaginary crimes of violence while, with what the Comtesse de Noailles called the "ennui characteristic of persons well born," we let real crime go undetected. And this in a country whose greatest danger lurks in the cynicism, class disaffection, and embittered sense of injustice that open our defenses to the gaudy promises of communism.

* * *

I have offered you a genealogical fantasy. This introduced the theme that medicine has much to offer but that your attitudes to such offerings decide what is to happen. Now I venture to indulge in somewhat more general speculation.

Everything has its price tag. Medical science is no exception. It has its own internal or self-generated problems. If we may take the credit for lengthening the years a new-born baby may expect to live, from 35.5 years in 1789[4] to 66.7 in 1946,[5] then we must face the problems of an aging population, i.e., where a much larger proportion of the citizens is elderly—really elderly, as distinguished from Bernard Baruch's sliding definition of an old man as anyone who is fifteen years older than you are. The advances of medicine have also diminished the death rate, thus leaving people exposed to more illnesses in their added but not always ebullient years. We have, in effect, traded mortality for morbidity, acute disease for chronic disease. Among our emotions, outright primitive fears have given place to the anxieties of a tightly scheduled, frequently repressed, and intensely competitive civilization. Having improved the mortality rates of the heedless masses, we find overcrowding, poverty, squalor and malnutrition—so-called population problems—pressing them: breadlines and riots in the name of justice in streets where babies' funerals used to pass in the name of God. The sociologist Robert Lynd recalled that when he was a boy some forty years ago in a small Ohio town, the community comment on the death of a child was, "Isn't it sad that God has called for little Elizabeth Kelley!" Very pointedly he asked whether the comment today might not probably be, "Why didn't they call a specialist earlier?" That question seems full of meaning to me. Not only has knowledge become power but, in the minds of men, the Almighty's grave responsibility seems now transferred to the specialist. From among the ranks of mere man the specialist steps forward, interposed, as it would seem, between Divine Providence and its hitherto inscrutable decrees.

I would hold that the following is an inescapable se-

4 In New Hampshire and Massachusetts.
5 In the United States.

quence: Science brings knowledge, knowledge brings power, and power calls for responsibility. In our adaptive society, where there is no longer an accumulated and applicable tradition to guide us, this newly imposed responsibility takes on heroic proportions.

My impression from both indoctrination and experience is that man's first reaction to responsibility is to dodge it. I do not mean that my early indoctrination was toward evasion, but one of the earlier stories I was told about the human race concerned Adam's reaction to the consequences of having eaten the apple of knowledge. He passed the responsibility to Eve, saying, "The woman gave it me and I did eat." I never fully understood the real meaning of that story until Science, like Eve, got the blame for man's use of the apple—this time the atomic bomb. In ever-increasing measure, man, through research, interposes himself into the hitherto incomprehensible mysteries of natural law. The results are variable, but some of them remind me of Oscar Wilde's observation, "When you break the little laws the big laws begin to operate." Discover a drug that banishes exhaustion or dims pain and you may have merely stifled the most valuable signal warnings of approaching disaster. As Frank Cobb said, "You don't cool a room by breaking the thermometer." We need a tincture of foresight with all our inextinguishable curiosity and single-track enthusiasm for knowledge.

Now in the light of these godlike responsibilities which attend new powers, or, if you prefer, in the light of the new problems generated by man's recently acquired control of powerful forces, I remember from college days a wise remark by Professor Thomas Nixon Carver. He suggested to us that the history of the human race could be written in terms of the survival of adversity—adversity in three principal forms: first, diseases, especially epidemic diseases; second, famine or malnutrition; and third, tribal or international war. He noted that he was far from thinking that more wars were unlikely (the year was 1910) but he did believe that medicine had made decisive headway, especially in the control of epidemic disease, and that communications and transport had largely banished the specter of famine from most parts of the world. He then looked out over the class and said, "I think the problem of surviving adversity has changed into a different problem, for which the human race is but poorly prepared, either by experience or tradition. I suspect some of you young men will have to face it—that problem is the problem of surviving prosperity." There is, I think, a reasonable connection between what Carver meant by prosperity and what we mean by the power over nature that comes from research. And let me interject right here that I am not thinking merely of atomic power when I say power obtained through research. The mastery of atomic energy can be used for destruction as well as for producing kilowatt hours of electrical energy. But there are other kinds of powerful knowledge. A comparable increase in psychological knowledge would make possible more subtle and devastating interpersonal cruelty and abuse just as effectively as it could contribute to the grace, the skill, and the beauty of being alive in a world filled with other human beings. Again and always the responsibility rests mostly with the consumer. You are the consumers: the decision as to the use of knowledge is yours, not the scientists' alone. You can't evade or avoid or ignore it; indeed, falling asleep in the face of responsibility would be an opinion—to be repented at leisure, if that is what leisure is for.

Confronted, as I believe we already are and shall be increasingly, by increased knowledge, power, and responsibility, what can we do? I recall a desperate night in a storm off Cape Ann in a yawl that was manned by myself, my college roommate, and his Norwegian sailor. At the top of our lungs we asked Albert Olsen what he thought we ought to do. The reply came back through the gale, "Vell, Mr. Ayer, there ban no back door: I guess ve stick it out." So much for evasion, passing the buck, or avoiding the blame. One or two points, however, deserve to be kept in mind to hearten us.

First, let us keep thinking of the possibility that the qualities of the human stock could be as much improved by research in human genetics as the quality of our plants and animals has been improved by similar studies. Visionary? Yes, visionary, but, as I think, neither impracticable nor unreliable. Indeed, I would place more hope in the returns from research in human genetics than in any other single field of study aimed at the improvement of the human race, both in point of avoiding the perpetuation of evil and reaching out to the realization of beauty and peace. Because the leader's commands, the priest's exhortations, and the philosopher's advice all aim to improve the conduct of man, may we sensibly conclude that human conduct is determined by nothing else? Does not acceptance of advice depend in part on the characters of those who do the accepting, as does government on the consent of the governed? I know that what is merely fashionable affects human conduct, but I am more impressed by the fact that some kinds of people will accept one leader, or law, or religion, or set of values, while another kind of people prefers something quite different. It takes, for example, a wise man to recognize wisdom. As the logician Wolf observes, "People can and do reason correctly without the aid of Logic; if they could not do so, Logic itself would not exist. God (as Locke remarked) has not been so sparing to men as to make them barely two-legged creatures, leaving it to Logic to make them rational. Rather it is the native rationality of man that has made Logic itself possible."[6] I would agree, and in agreeing go as far as to suggest that a similar statement could be made as to a number of other admirable and native human traits besides rationality. If, as C. C. Little has shown, different strains of mice—one, for example, consistently quarrelsome and aggressive and another peaceful and submissive—can be bred out of a mixed population, and such strains breed true, we may be justified in wondering what kind of people are most disposed to accept an offered credo and make it work. Only as good people multiply on this earth can goodness expect to interest, attract, convince, and reward successive generations.

Next, let us remember amid our increasing responsibilities that an organism is an assemblage of organs and tissues so intimately interrelated that no marked change can come to any one of them without affecting all the others. Should not that fact set us to wondering whether so much separation of special knowledge among isolated specialists is safe? Also, most higher organisms possess special senses such as sight and hearing whose function is to adapt and adjust the organism to changes in its environment. In the natural sciences, the medical sciences, yes, and in the social sciences, we need more of the ecological approach, the holistic view. Such words, and the ideas they represent, should be commonplaces—but they are not. In simpler terms, we need a more comprehensive understanding of the whole picture of our lives, of relationships as well as of things, of context and the attendant circumstances. We shall not see life steady until we see it whole. It is said that women are wiser than men because though

[6] A. Wolf, *Textbook of Logic* (Second Edition, London: George Allen & Unwin, 1948), p. 35.

they know less they understand more. I am not pleading for mere erudition or glib versatility or existentialism, but rather in the vein of the biblical warning, "with all thy getting, get understanding." We could well ask ourselves before our civilization falls to pieces—specialized pieces, of course, and each highly honorable and praiseworthy but hopelessly isolated from each other—whether it isn't high time to give enthusiastic allegiance to the kind of minds that can encompass and comprehend the essentials of many branches of knowledge. Is it not reasonable by now for a university to maintain just one chair for a professor chosen for the breadth of his understanding of relationships when so many other chairs are accorded on the basis of depth of special knowledge? God knows, we want such understanding in our university presidents. But where to train it? For the responsibilities consequent to knowledge and power are not merely deeper: they are broader responsibilities than ever before.

Another approach may help us to become answerable for the increasing powers medical science will bring. I would call this approach "naturalism." By "naturalism" I mean that we should realize that we are what we are as a result of thousands of years of selection and survival, of trial and error, in short, of past success—success it is true, but then in an environment quite different from the environment we now create for ourselves. Just because we have been the winners in one kind of game, can we expect to win under a different set of rules or conditions? If we are the survivors, and so the competent product, of tens of thousands of years of plenty of muscular exercise, of plenty of sleep, of much coarse and bulky food, of much leisure and play, of hardship, too, and of a constant input of different sensory impressions though a relatively simpler emotional life, it is at our peril that we increasingly curtail physical exercise, substitute drugs and electric lights for sleep, choose concentrated foods, cheat ourselves and our children of leisure time, avoid the exercise of tenacity, endurance, and fortitude, minimize our sensory impressions and complicate our emotions by conflicting loyalties in a relentlessly exaggerated competitive existence. Human evolution never faced such a summation of new circumstances before. Our minds and bodies evolved in quite different conditions. To such brash and unnatural overloads I doubt if we can add indefinitely the immense new responsibilities of unfamiliar power. We should remember that in one sense we are very old to take on so many new conditions of living. If we take up heedlessly with every new thing that comes along, we shall deserve Will Rogers' delightful reproof, "They is a lot of monkey in us: you can toss practically anything you want into our cage and we'll give it serious consideration." Or Mrs. Wharton's neat reference to "getting into the thick of thin things." I am not pleading for a hypochondriac hygiene or for a furtive timidity about living but rather for the preservation of an honest sense of when in each month's living we have had enough of all that is new to the human race. As Walter Cannon's *diener* used to say, "There's no whistling without your upper lip." There is something fixed and immobile, tried and true, that comes to us from our biological past. Let us take that into our reckonings and be willing at times to keep our hair shirts on.

Shall we now leave philosophizing to ask ourselves the individual question, "What is my attitude toward the knowledge and practice of medicine? What can I do about medicine—its furtherance and its application?" I don't believe your conduct will change at all unless you get a new view of yourself. Here the Chinese proverb is apposite, "Great men never feel great: small men never feel small."

Throughout this talk I have laid stress not so much on facts, dates, and events in medicine as upon attitudes and value judgments. I hold quite simply that you need feel neither great nor small: all that is needed is that you participate. What will happen in research, in teaching, in medical practice will chiefly depend on your layman's attitude—on what you think of unnecessary pain, physical and mental, of decrepitude and death. What price do you set on health—yours and others'—particularly others'? For you will also have to decide what you want to do about cripples and defectives in our highly competitive society. My brother once, in exasperation with a stupid ditch digger, exclaimed, "Haven't you any brains!" The workman from his ditch replied, "If I had brains I might be where you are and somebody else would be where I am," a remark which left the essential problem just where it ruefully belongs. What action you take will depend on your own weighting of the relative claims of the common welfare versus your own personal competitive success; in other words, on your attitude to yourself as related to persons less favored than yourself. One needs some convictions in such matters to organize the application of medical science to the needs of mankind. One must know the *why* as well as the *how* of using medical science.

I hope you aren't one of those who have to be reassured. I am thinking of the young man with an introduction to the Irish novelist George Moore who asked, "Mr. Moore, do you think I should go in for a literary career?" Moore promptly replied, "No." A bit nettled, the young man expostulated, "But Mr. Moore, you went in for writing when you were young and you have been successful." "Yes," said Moore, "but I never had to ask anyone whether I should or not." You see, it depends in large measure on attitude.

But whether you do research work or support it, whether

you teach medicine or help it with even a hundred-dollar cheque written out tonight (or for five hundred dollars in the morning when your head is clearer), or whether you help the care of the sick in any way you please, I beg of you, *don't* remain a spectator. With the movies, the radio, grandstand sports, the newspapers, the commentators and the columnists, and the captious critics of a government they wouldn't serve, we have quite enough spectators—and not enough good players of anything more important than a saxophone. Without more good players for the harder game of being responsible for our new powers, we cannot meet Whitehead's definition of progress: "The art of progress is to preserve order amid change and to preserve change amid order."[7] That is the essence of progress in medicine, too. We have need of reordering our teaching in the light of changes imposed by new knowledge; and we have need of change amid the order imposed by professional imperturbability and intransigence. We doctors can't do it all—we are too intimately involved to be dispassionate judges of what should be done. Napoleon is said to have observed that if you want something well done, get an expert to do it and a layman to see whether it has been well done or not. The latter is your job, whether you like it or not.

I come to the end. Because Benjamin Franklin characteristically looked forward, let me suggest that historical perspective is a term that can apply to the future as well as the past. Imagine if you can a dinner party given in Philadelphia in the year 2375, again a dinner party of twelve men, but this time with you as the strange ancestral codger at the left of your host. All those present would be bearing your name or at least the inevitable fraction of your genes. If knowledge advances, as I expect it will, you would be

[7] A. N. Whitehead, *Process and Reality* (Macmillan, 1929), p. 515.

more astonished and bemused at the conversation than any of your descendants around the table.

For I can end this essay with no panegyric upon the omniscience of modern knowledge but rather with cheerful resignation based upon the fact that it is of the essence of medical science to make our present knowledge seem incomplete and quaint just as soon as can be. Indeed, I have an even more important conviction: the true outwardness of medicine, that is, our attitude toward medical knowledge, will change too, and for the better.

WENDELL H. CAMP

Plant Sciences in the World Today

It is my avowed purpose to take the view that man, currently with weapons of terrible destruction in his hands and his judgment often warped by fear, is not today standing at the end of his biological line, or even at a mid-point. As a biologist first, and only by specialization a student of the plant sciences, it is my fundamental belief that man as yet has made only scant use of the latent potentialities for his own physical development and cultural betterment.

But let us be neither deceived nor deluded. Necessity is not always the mother of invention; to maintain and develop a civilization it is needed that we also look ahead and be prepared. In these perilous times when the bases of civilization seem almost to be threatened, let us remember that the way to salvation lies not in blindly following anyone who promises to save us, either from our enemies or from ourselves, or in any panaceas which may be propounded, but in a calm, objective view of the situation, making certain at all times of a constant vigilance against the little things which corrupt and undermine our basic culture. With this vigilance must be a continued diligence of application of such knowledge and skills as are at our command. And, in addition, if our culture is to move in a forward direction, the human mind must at all times be free to venture into new channels of thought.

Needful as it is to be deeply concerned over present world conditions, we must not become blind to the basic situations which ultimately will decide whether or not our own cultural pattern is to develop along increasingly more satisfactory lines. While it may be necessary for us to defend ourselves with our right arms, fear must not paralyze the left, for they must be equally busy with the problems of the reinforcement and advancement of our civilization. Even without alarms and the wastage of wars, civilization sometimes sows within itself the seeds of its own destruction. Although unpleasant to contemplate, it would seem best that we understand their sprouting, the nature of their growth, and the means of plucking them out before they come to monstrous fruition.

THE PRODUCTION AND ENLARGEMENT OF ACID AREAS

In the third lecture of this series, Dr. Noyes recalled to us that the diameter of the Sahara Desert was enlarging at the rate of about a kilometer per year, and that checking or reversing this trend was one of the things which challenged the ingenuity of the scientist. I am not personally conversant with the situation on the edge of the Sahara. However, some years ago in southern Mexico I chanced to be on an eminence where I could watch the course of a moving thunderstorm. As it passed from a forested region across a cleared and cultivated area, the rain suddenly ceased falling, and did not begin again while I watched the course of the potential rain cloud as it moved on beyond the cleared area, over the forests, and finally out of sight. I became curious, and in subsequent years at other places in the tropics and semitropics I have noted the same phenomenon. It also occurs from time to time in temperate regions.

The explanation, I think, is quite simple, for recently I had good proof of what takes place while in a low-flying plane going over a certain fairly well wooded—and, I should add, comparatively flat—area in New England. The day was sunny and warm and, as we passed over the forested area and came to a cultivated plot, almost invariably there would be a bounce caused by a sudden up-draft of air. Apparently the foliage of the trees did not reflect the light in the form of heat, or at least not in sufficient quantity to cause any great amount of air turbulence. However, each treeless, cultivated area seemed to act as a heat generator, resulting in a column of warm air which constantly rose to a considerable height over each of the cleared plots.

In Mexico, where I had first noted the foregoing effect on a passing shower, apparently such a column was rising from the cleared and, at that period of the year, bare land, its air sufficiently warm to disperse the droplets of incipient rain and maintain them in vapor form within the cloud. Furthermore, as I have usually noted since, it is some distance beyond the point of contact between the cloud and the rising column of warm air before the cloud has cooled sufficiently to begin the precipitation cycle again. Such a phenomenon is not likely to be noted during periods of larger epicontinental storms; its effect seems to be most pronounced on the local showers during otherwise hot and sunny weather.

Under conditions of climatic or edaphic tension where the showers and thunderstorms of the drier season make the difference between a forested area and one of grassland or even semidesert, the loss of these beneficial showers—

far beyond the original up-drafts from the clearings that seemingly stop them—would soon lead to a drastic change in vegetation cover. Thus, where man clears land for agriculture in climatic or edaphic tension zones he is not only likely to curtail seriously his cultivated area, but also to destroy an ever widening zone of forest, a thing he never intended to do in the first place. I am increasingly of the opinion that a considerable portion of the useless, semi-desert thorn scrub of the American tropics and subtropics developed in just this manner. I cannot extend this to the rest of the world from my personal observations, but a perusal of the literature on vegetation changes following the advent of man into certain areas leads me to wonder if here we do not have a general hypothesis which should be tested with some care. If this hypothesis proves correct, it would mean that we should be most careful as to the type of ground covers we supply and the seasonal cycles of cultivations where there is danger of this secondary and certainly unwanted effect.

In this connection I have lately become curious about the situation in the so-called "dry farming" areas of our Great Plains region, where the land is maintained as bare and continually cultivated, weed-free fallow for a year or more between crops. Does this practice lead to a continued and progressive diminution of precipitation on these areas relative to what it would have been had they been left with their natural cover of grasses? If so, then we are not only in danger of soon losing them as grain-producing areas, but also of making them unfit for future grazing of meat-producing animals. We must not forget that there is a soil moisture quotient below which even the hardier of the plains grasses will not exist and cannot be successfully reëstablished.

From the abundant archaeological remains we may be

certain that man once was quite abundant over large areas of what now is the Sahara Desert; he also once carried on a considerable amount of agriculture there. One therefore might perhaps wonder whether the present Sahara Desert, with its obviously ever-enlarging rim, is entirely a product of natural climatic desiccation, or whether man might not have been a rather large factor in the initiation of this now self-perpetuating arid waste.

Let me hasten to say that I am aware of the correlation between wind and ocean current patterns, geomorphic features, and the world's great scrubland and desert areas. But I am equally aware that the vegetation of certain other areas is anomalous and does not seem to follow the expected pattern as outlined in our textbooks. Although the working hypothesis set up here applies primarily to climatic tension zones, there are known examples where once mesic, cultivated areas have changed to semiarid or even desert areas within historical times. And let us not forget, since Pekin Man is known to have used fire, that man was clearing the land for one purpose or another, with total-destructive fire his usual tool, not less than nine hundred thousand years before the dawn of recorded history.

Those of us who have cleared and tended plantations in the rainy tropics know what a never-ending task it is to keep the jungle pushed back and the cultivated crops free of choking weeds. There is little time for the enjoyment of leisure or the development of a high level of culture. Man long ago learned that the ideal place to carry on agriculture was in the zone between the rain forest and either the steppe or the semidesert. It is in such areas—actually on the dryish side of mesic conditions, where periods of work in the fields are followed by harvest and sufficient time for contemplation and development of the arts—that the world's great civilizations have arisen; it is here in these climatic tension zones, also, that they have all too often brought about their own degradation through careless use of the land.

THE PROBLEM OF AVAILABLE CROPLAND

According to the best estimates, of the earth's sixteen billion habitable acres only about one-fourth can be classed as producing cropland. These four billion acres must be apportioned among a population of something over two billion. Thus there are slightly less than two acres of cropland per person.

Nutritionists inform us that, for an adequtely balanced diet, two and one-half acres of cropland are necessary per person. Therefore it is immediately obvious that a considerable proportion of the world's people are today undernourished. It is very likely that this as much as anything is at the bottom of the world's present unrest.

Malnutrition is not understood by everyone. It does not necessarily consist of a lack of basic calories. As one travels about among various peoples and studies their food habits, it becomes obvious that one of the first effects of insufficient cropland is to cause a shift from a balanced diet toward the more concentrated carbohydrate foods. And various of the basic carbohydrate calorie producers are notoriously low in necessary vitamins and often deficient in vital minerals. The first result of this shift in diet is a decline in mental alertness and physical vigor.

When the foregoing statement was made some time ago, it was soon challenged by various people. Toward the end of 1948, in an issue of *Lancet,* the British medical journal, there appeared an article which bore out much of this, and with striking clarity.

Some of those recently returned from England have noted in the public press that, although beset with numerous restrictions and forced to live on an "austerity diet," the British appear to be quite well fed. Dr. Instone's report in *Lancet* is by no means a full one, nor does it cover the whole field; but it is an indication of what is happening in England at the present time.

On their present austerity diet which, as noted, would be shifted toward the concentrated carbohydrate side (as it has in England), many have gained weight. Therefore in general appearance—and especially to the untrained observer—the British seem to be adequately fed. However, the majority of those examined complained of being perpetually tired, in spite of the fact that they had advanced their sleeping time to as much as nine and a half hours per day. Colds and minor infections were on the increase. Many of those examined showed marked symptoms of mental confusion, lack of alertness or, worse yet, carried with them a continual feeling of frustration and defeat. These are the first symptoms of vital nutritional deficiencies. It is not necessary to philosophize on the eventual outcome of this situation if the present dietary conditions in Britain are permitted to continue. It is therefore necessary that we take cognizance of our situation here in America and assess our basic food resources.

When the first settlers of our own era stepped ashore in America, something like 560 million acres of excellent, potential cropland lay before them.[1] By 1918 the population had increased to the point that each person in the United States had only four acres of cropland at his disposal. Today, the figure stands at just about 3.2 acres per

[1] Recent estimates of a billion acres of agricultural land available in the United States are misleading, for they include large areas suited to controlled grazing but basically unfit for agricultural crops. This figure also includes a considerable area of submarginal land which at one time or another has been cultivated, but which should be classed primarily as potential forest or grazing land.

person. At the present time the population of the United States is increasing at the rate of 2,200,000 per year. This would seem to mean that every year about 5,500,000 acres of our present working margin of available cropland must be turned from the production of fibers, oils, and other important industrial crops into food production if the present level of nutrition of the American people is to be maintained. Furthermore, it does not take any particularly complicated mathematics to discover, if the present trend keeps up—as it very likely will, and if we find no other way out—that in another thirty years, or about 1980, the American people will have reached the point where there will be only the requisite 2.5 acres of food-producing land per capita. What lies beyond that point might easily be conjectured if one has any sense of history.

Of the original 560 million acres of cropland, already about 100 million acres appear to have been lost through erosion and misuse. And if adequate steps are not immediately taken, by 1960 another 100 million acres will have been discarded. Therefore it is obvious that, if our land-use pattern is not immediately modified, and if we make no advancement in our crops and cropping methods, in another thirty years there will be only enough cropland left in the United States to feed some 100 million people. Yet with our present increase in population it seems likely that by 1980 there will be some 175 million mouths to feed. Thus, unless immediate and heroic measures are taken to save the remaining soil we have, and also some other way found to alleviate the situation, the critical point of 2.5 acres of cropland per capita will be reached, not some thirty years from now, but in the next fifteen or twenty years.

Industrialists, especially those whose manufacturing processes are based on such things as vegetable fibers, oils, and similar products, should be prepared to face this situation. We cannot borrow from Peter to pay Paul; to maintain physical vigor and mental alertness in the American people there may well come a time when we dare not use cropland to produce non-food crops. And here is where the pinch will come.

A civilization cannot exist on food alone; other plant products must be considered. In addition to the 2.5 acres per capita thought to be needed to maintain a minimally adequate diet, at least another half-acre of cropland is needed for the growing of industrial materials such as fibers, oils, and similar products.

Wood also is required, but this need not be subtracted from the cropland area since the forests can be grown on land not suited to other crops. However, we must face the fact that today in the United States we actually are using wood about twice as fast as we are growing it. A continued policy of debit forestry can lead only to bankruptcy in this basic plant resource.

Being the best-fed, best-clothed and most adequately housed people in the world, we somehow have the feeling that nothing can happen to change the situation. Actually, we are potentially very close to the brink of disaster, and it is very likely that our actions in the next decade will decide whether America is to continue strong and able, or whether it will start down that long trail from which no people—no civilization—has ever returned. Our working margin of cropland, the very basis of our civilization, is much too narrow for comfort. And it is continually shrinking.

At this point, lest you think I am one of those who go about crying that all is lost, let me say that, dark as this over-all picture may seem, it certainly is not a hopeless one. With immediate action we can go quite a way in saving much of our remaining 460 million acres of cropland. We can even rehabilitate some of that which now is discarded. A vigorous program of reforestation must be begun on a scale which few have envisioned. And lastly, the whole force of our knowledge of the living plant must be brought into play on the problem of the production of important plant materials. It is not enough that we have prolonged the life of the individual through advances in medical science; to be an integrated and functional member of society this individual must also have a balanced diet, be adequately clothed, and comfortably housed. And all this, in large measure, depends on plants or products derived from plants.

THE WAY AHEAD

It is obvious that if nothing were done to increase the productivity of our crops, the population would soon overtake the available cropland and it would then be necessary to devote all our cultivable area to the production of food. This would leave a considerable portion of our industries —all those in any manner dependent on products from cropland—in a very bad way, and would soon throw our entire economy out of joint. There is no use saying that food is more essential than such things as cotton or linseed oil; all these are necessary and a part of the total picture of our civilization. Therefore, with our rapidly dwindling margin of cropland, it is vital that every effort be made to put our plant materials to the most efficient use. Various lines of attack are open; they should be parts of a common solution to the manifold problems that confront us.

PLANT BREEDING. Better and more efficient plants must be produced. Although plant breeding and selection (particularly the latter) have been practiced for centuries, it is only within the last few years that we have learned really to unlock the hidden hereditary potentials in plants; in some instances we have produced not only new and better varieties of plants, but have actually synthesized new species, thereby opening up a whole new vista of possibilities. This work must be pushed at all costs, not only in food plants, but also in those yielding industrial materials, as well as among the forest trees.

The recent 25 per cent increase in the production of corn on the same acreage as used previously, brought about through the use of the new "hybrid" varieties, is only a beginning. But it is an indication of what may yet be accomplished in other plant materials. The recent synthesis of new species of grass yielding more than 100 per cent increase in forage, and at the same time capable of competing with and exterminating the poorer-producing native range grasses, is another instance of what the plant breeder, working with modern techniques, can do. Likewise the breeder of forest trees has recently been able to produce some hybrids which promise to yield quality lumber in one-third to one-half the usual time. This automatically would cut down the acreage needed to produce wood.

One of the things confronting the biologist is the variability of organisms. There was a time, not so far distant in the past, when biologists were not concerned with variability; if they considered it at all, they treated it most casually. I do not feel that it is being unduly mechanistic to assume that the bulk of variability of reaction in organisms—be they men or maples—is the result of genetic differences. It is a simple statement of the facts as we are discovering them in our routine daily researches. It would be most unjust to say that the layman does not understand the nature and extent of this variability if by so doing I were to imply that all biologists did. The fact is that biologists themselves are often unaware of its qualities and magnitude.

As an example of the variability which the plant breeder encounters and is studying so that he may control it and thereby produce more efficient organisms for our gardens, fields, and forests, we might briefly consider the inheritance of flower color in the common garden dahlia. As we are now all aware, genes are the controlling catalysts of hereditary differences. It is not mere chance that the garden dahlia comes in a wide variety of colors and shades. Two groups of wild species occur. The first of these, consisting of about a half-dozen species, is characterized by varying tints of magenta anthocyan and ivory flavone pigments; the other group, of about four species, is known to contain scarlet anthocyan and chalkone yellow. The first group appears also to contain a special gene for the inhibition of the production of the ivory flavone pigment, and the second group definitely does have a gene set progressively inhibiting the production of the yellow chalkone.

Our common garden dahlia is not known in the wild; it was already in the gardens of the Mexicans when the Spaniards arrived. Apparently it had appeared in those ancient gardens through progressive natural hybridization of the species during that period when the Mexicans used the dahlia root for food.[2] The superior qualities of these

[2] The early Mexicans domesticated various of the wild species of dahlia, not for their striking flowers, but for their nutritious tuberous roots. Although Mexico does have its indigenous tuberous potatoes, per unit of food producing area they are not nearly so productive as the dahlia. The potatoes now raised by the Mexicans appear to have been introduced from South America by the Spaniards some time after the Conquest. In fact, if one dare read between the lines of the Popul-Vuh, the sacred book of the Maya-Quiche, the Mexicans may have got their present staple crop, corn, rather late in their development. It may well be that in Mexico—for this was not the ancestral home of corn—the dahlia has been cultivated for a longer period even than corn.

hybrids were noted, so that, by the advent of the Spaniards, few if any of the wild species types were still in cultivation. Now it is not my intention to bore you with the details of the inheritance of color in the garden dahlia, for there are highly complicating factors, only one of which is that the plant is an octoploid, that is, instead of having the normal two sets of chromosomes, it has eight full sets of chromosomes in each cell. Be that as it may, by a set of calculations it may be determined that, from these relatively few sets of colors, the breeder of dahlias could produce no less than 134,217,728 different and definable colors, shades, and tones in the flowers of this interesting garden plant. You will now begin to understand what we mean when we say that the plant breeder has only begun to tap the variability inherent in the materials at his command.

But, perhaps, with the fiscal expenditures of Government soaring to points far beyond such a figure so that it now seems almost puny, let us delve further into the nature of this inherited variability. It is most difficult for the geneticist to keep track of these elusive genes unless they come in contrasting pairs. Fortunately many of them do, and so we have methods of following the individual genes through successive generations of offspring. One of each pair of contrasting factors is technically known as an allele. Thus the gene determining the dwarf (bush) form of the garden pea is an allele of that determining the tall-growing form, as the gene determining white flowers is an allele of that determining blue flowers, and so on.

No one as yet has satisfactorily determined the number of genes involved in the production of an organism; it may be a thousand, it probably is thousands. So, merely to get on with our problem, let us make it simple and suppose that an organism has only 100 sets of genes in its hereditary complement; let us also suppose that ten of these possess mutant allele mates which can be readily identified. Let us further postulate that there is free combination between these various gene groups. The question then would be: How many different hereditary gene combinations might the geneticist expect to uncover in the course of his work on such an organism? I will not answer this in a simple figure involving only a hundred million or so, as was done in the case of flower color in the dahlia; instead I will use mathematical shorthand and say that the number would be in the nature of 10^{100}. To the uninitiated, 10^{100} may seem like a fairly simple number, but the mathematician will tell you that, as a real number, it cannot be comprehended by human consciousness. The nearest we can come to understanding it is to remember that, some years ago, Professor Eddington calculated that the number of electrons and protons in the matter of the entire visible universe is no more than 10^{79}.

Chemists and physicists are prone to dwell at length on the complications they encounter in their work. Usually they begin moaning and making other queer sounds when they are forced to deal with as many as three variable factors in any problem. At the same time, off in a corner of his breeding plots, the plant geneticist often is struggling with as many as twenty-five identified pairs of variables; in the corn plant the known number is considerably larger, and the number of unidentified pairs may be legion. It is therefore only legitimate to ask how it is that the plant breeder ever finds his way about in this maze of gene material. Let me assure you that it is no easy task, yet this is scarcely the place to expound the method. I will, however, say that the information as yet is not on the "classified" list; it is still open to anyone in the world with time and patience enough to sit down with the literature and dig it out. I pray that it may always remain so, for this information, if diligently pursued and properly applied, will *not* be a means of degrading and ultimately destroying man, but of saving him and his civilization.

Yet, as ever, a little knowledge is a dangerous thing. There are many who seem to think that at last we have here a ready panacea for our ills. We are all aware of what has been done with various crop plants in recent years, especially with the corn plant. I do not decry the advent of hybrid corn; it was a necessary development along the only course open to us. Yet it is not an unmitigated blessing. The thing we seem to have forgotten in all this fuss and hurrah over hybrid corn is that the first settlers often harvested as much corn per acre with their inferior sorts as we are at the present time with our highly specialized hybrids. In a former passage I noted that hybrid corn, now in production on about 70 per cent of the corn acreage in the United States, had raised the level of production in this one grain about 25 per cent. I now must point out that this increase from the same acreage is relative only to those last years before the cultivation of hybrid sorts became general. This will indicate something of the serious depletion of soil fertility which, like a slow and creeping paralysis, has been overtaking the level of productivity of our best mid-continent croplands, an agricultural area which our fathers boasted possessed a soil that was inexhaustible.

There is no doubt that modern hybrid corn is a more efficient producer of plant material than the older varieties. But it is time that we determine the nature of this efficiency and understand what it portends for us. The farmers of the corn belt are discovering, somewhat to their chagrin, that they cannot raise this new hybrid corn in the same old ways and maintain its high level of productivity. The increase of the first few years with the old methods was a snare and delusion, for this increase was the result, primarily, of the hybrid's better ability to extract the necessary minerals from the soil. But like any mine, there is a limit to the minerals which may be taken from it. Just as mines are worked out more quickly with modern machinery, so does this new hybrid corn all too rapidly deplete the mineral supplies of the fields where it is grown. As a result, to hold the corn belt to its maximum production, it will be necessary for the agriculturists operating with this new type of corn to enter into a fertilizer program which they had not contemplated. They are learning the hard way that they can't get something for nothing.

Today American agriculture is at a definite turning point. For too long we have gone under the supposition that if we plowed the land, planted the seed, and kept the weeds down, the plants would do all the rest. Sometimes we threw on a little manure for good luck. A glance at records of the amounts and kinds of commercial fertilizers bought by American farmers during the last half-century indicates that it is only within the last few decades that they have had any adequate realization of the needs of balanced mineral nutrients. In the growing of crop plants, the time is come when we shall have to consider the soil primarily as a medium which holds the plant more or less erect so that the photosynthetic surfaces, usually the leaves, are best displayed to the kinetic action of the sun's rays. Such mineral nutrients as the plant needs to operate at maximum efficiency must be supplied at the proper times during the plant's growth, and not in a desultory fashion as something we can expect the plant to find for itself. When this viewpoint is understood by every tiller of the soil, American agriculture can then be said to be on a sound basis. It is not enough that the plant breeder supply the farmer with more efficient plants; the conditions for these plants to be maximally productive also must be provided. The duality of the process must be understood.

MINERALS, PLANTS AND MAN. Here is a topic which, as a plantsman, I should like to dwell upon, not for a few brief moments, but for a whole evening.

I do not as yet have the 1950 figures, but in 1949 the farmers of the United States used something over 16.5 million tons of commercial fertilizer. Only a decade previously, they were putting less than half this amount on their fields. The cost of these fertilizers to the American farmer in 1949 was about 600 million dollars. There is no way of accurately estimating the profit on this investment in increased crops, for at the same time there have been considerable advances in methods of tillage, wider application of adequate land-use practices, better farm machinery has been developed, and new methods of disease and pest control brought forward—in addition to the constant introduction of new and improved varieties and hybrids. But the increase, as a result of additional use of fertilizers, must be considerable.

As noted earlier, the advanced cultivator is beginning to think of the soil primarily as a medium for holding the plant erect, and that it is his business to supply it with the requisite chemical elements for it to function as a well-integrated and smoothly running synthetic machine. This is all to the good. Also in earlier passages I seemed to be gnashing my teeth at the loss of much of our potential cropland through misuse. It is greatly to be deplored, for much of it is now unfit for the growth of plants. However, with adequate measures, when coupled with our knowledge of plant nutrition, certain parts of these despoiled acres can again be brought into production—provided we proceed on the assumption that we shall have to supply almost the entire amount of growth-producing minerals. Here is a field that needs a great deal of further investigation, for such soils often are in such poor physical condition that they require very special treatment before plants will grow there, even if the necessary mineral elements are supplied.

Also, we are just discovering that the mineral balance in the soil directly affects the health of the individuals consuming the food produced on them. This is a considerable topic in itself. However, I cannot emphasize enough that here is a problem that needs our immediate attention. Many of these secondary effects are as yet quite obscure; there are indications that they may be part of the recently recognized trace-element situation, but other factors indicate that this is not the whole story. I will leave this interesting and yet little understood subject with the note that, because of some soil situation as yet unknown, the green vegetables from certain parts of our truck-farming areas appear to be inadequate suppliers of the anti-pellagra factor. As a result, although people using vegetables from these areas ought to be getting quite enough of this important substance, they yet have an incipient form of pellagra. Usually it is not sufficiently severe to exhibit the advanced symptoms known to the medical profession, but enough to cause secondary debilitating effects in the person so afflicted. The medical profession is not as yet generally aware of the situation, although some have discovered that niacin treatment relieves the condition almost immediately.

Before I close this topic it should be pointed out that, if we were to apply the requisite mineral nutrients on all soils and for all crops that need them, the cost would be almost prohibitive. Again we would seem to be reaching an impasse, with no way out of our dilemma. Happily, there is a way. Recently we have discovered that, with some of the more expensive fertilizers, it is not necessary to waste them by burying them in the soil, where they can be reacted upon by a wide variety of substances, often thrown into insoluble forms, or leached out by rains, and so lost to the plant. Instead, to our slight amazement—for the plant scientist can yet be amazed at the extent of his ignorance—we have discovered that many crop plants can absorb a remarkable amount of mineral nutrients through their leaves. These materials therefore are sprayed on the leaves at the exact time in the plant's development when a particular element—or combination of elements—is required, and only in the amount needed to produce a particular effect. This not only leads to a really great saving in the amount of fertilizer used, but also permits a much greater integration of the whole synthetic system of the plant so that the agriculturist now can actually control the production of a particular crop, much in the manner that an industrial chemist controls the synthesis of his manufactured product. With certain crops, we now dare begin talking about an engineered agricultural product.

OVERLY REFINED FOODS. The question of food-processing would seem to be adequately cared for, yet quite recently a chemist informed me that much of the food reaching American tables was so "refined" that little but its caloric value was left; he added that the addition of accessory vitamins and minerals, as now widely practiced, did not begin to return the whole nutritive value of the original food material. Being the chief chemist for a large and well-known concern processing both human foods and animal feeds, he was in position to say with complete authority that much of the really nutritive material actually was refined out of the foods destined for human consumption and put into the foods destined for animals and poultry. He was entirely serious when he remarked that

"On the whole, American hogs today are fed a better balanced and more efficient diet than the American people." It therefore would seem that there is considerable room for the education of the American people in what constitutes a really nourishing and economic diet; once demanded of them, the food processors would automatically fall into line.

TECHNOLOGICAL ADVANCES IN INDUSTRY. A long-range industrial research program must be set up which will be designed to make the most efficient use of basic plant products. It also should ramify into the field of substitute materials. For example, the individual cotton grower may raise a great clamor over the increased production of synthetic fibers; actually, every ton of synthetic fiber produced (if obtained from some substance not also produced on cropland) actually releases an equivalent acreage for the production of food, or for some industrial plant product for which no adequate substitute has yet been found. The same principle can be extended greatly, when we once fully understand the need for such a program of research.

WASTE OF PLANT SUBSTANCES. Waste of materials produced by plants must be drastically cut. Although large strides have been made in this direction, there yet is considerable waste right on the farm, with much material lost which might be used in industry or, if properly processed, might be returned to the land in the form of effective fertilizer. Also, plant-product wastes in certain industrial processes are so high as to be a scandal. It cannot be said with enough force, but every bit of such waste represents an equivalent loss of valuable productive land.

The tally sheet is clear: Today, each person in the United States has at his disposal 3.2 acres of cropland. Of this, 2.5 acres must be devoted to food production, and an additional 0.5 acre to the production of industrial crops. Therefore our present working margin is only 0.2 acre per capita. And with the present increase in population this working margin of land could easily be used up in the next few years. The production of more efficient crop-plants is not enough; we also must eliminate every bit of wastage of vital plant materials.

FORESTS AS CROPS. As noted previously, already here in the United States we have lumbered off about four-fifths of our original forest cover. This is not so bad as it sounds, since it is not necessary to maintain all our original forest area to supply current and future wood demands. Yet we must not forget that today we are cutting our forests about twice as fast as we are growing them, with the greatest dearth beginning to be felt in the slower-growing hardwoods.

There are some who have advocated that we increasingly turn to our neighbor on the north for an additional wood supply. Although about one-third of the area of Canada is forest-covered, only about 700,000 square miles of this area can be said to be capable of producing continuous timber crops. Of this, only some 470,000 square miles can be considered geographically and economically suitable to timber cropping at the present time. Already pulpwood is being dragged from the Hudson Bay drainage, across the Height-of-Land, and floated downstream to the mills near or on the St. Lawrence River.

It has recently been realized by our Canadian friends that, like ours, their forests are not inexhaustible. At the present time the curves of annual consumption and regrowth of the Canadian forests are beginning to approach equality. With increased consumption this will soon be thrown into disparity and so constitute a danger to this source of supply. Europe is now over-cutting its forest increment by about 20 per cent, even in the great wood-producing Scandinavian countries, and Asia's forest resources have been in serious imbalance for centuries.

Let us not think of our forests only as a source of wood for construction and the arts. No civilization today can last unless there is an ample supply of paper for the dissemination of information and the free exchange of ideas. This business of civilization has many ramifications, yet I doubt if any will dispute greatly when I say that our books, magazines, and newspapers are the single greatest factor in the rapid spread of today's culture. Without free communication between peoples our civilization would start to decline at a perilous rate, and we should soon find ourselves back in an age of intellectual darkness. It is not the spoken word, but the written word that gives continuity and a necessary driving impetus to our culture. When, for any reason, our printing presses become still, that day will mark the end of civilization as we know it.

Today, with technological advances on every hand, our forests fulfill another function. Wood cellulose is a very useful material. From it the chemist can fashion excellent textiles, a great many of the plastics which enter into our daily lives are made from it, and numberless articles are now being developed in which this material is used.

It is therefore obvious, if we are to have a constant supply of forest materials for wood, for book paper and newsprint, and for industrial cellulose, that we shall have to rid ourselves of the idea that forests take care of themselves; our forests will have to be treated as crops and handled with the same care and skill as our grains and vegetables.[3]

[3] One of the problems of establishing and cultivating forest crops is that the majority of trees have a mycorrhizal associate, that is, there is a fungus which penetrates the roots and also ramifies through the soil. In fact, many of our trees cease producing the minute absorbing root hair cells, often thought to be characteristic of all roots, soon after passing out

For much too long we have been lumbering the land and letting chance determine what sort of forest replaced the one cut down. We have been exceedingly careless about this matter. It all goes back primarily to an old lack of understanding of the nature of the inheritance of variation in nature. For example, at the present time we are now in our third or even fourth cycle of cutting of the white pine stands of New England. In certain areas foresters are beginning to note the increased incidence of misshapen and undesirable (cull) trees. Usually this is thought to be an effect of storms and hurricanes.

If we stop to consider how our reforestation has been carried on in the past we shall see the fallacy of this view. In each previous cutting cycle, only the best of the trees were felled, leaving the gnarled, misshapen individuals,

---

of the seedling stage. Thus all the water required by the tree first passes through this mat of fungus that sheathes the tree roots. Likewise, all the mineral nutrients required by the tree also are transmitted to it through the fungus; in fact, there now is good evidence that a large number of our trees are not successful unless the fungus first works over these nutrients and converts them into forms more easily assimilated by the tree. In return for these services, the tree supplies certain basic food substances and "vitamins" (growth regulating hormones) needed in the development of the fungus. Thus, in establishing a successful forest, it is required that we also consider the conditions necessary for the maintenance of healthy mycorrhizal fungi. The silviculturist is not only growing trees but is also culturing the fungi on which his trees to a large extent depend. In garden practice a "balled and burlapped" (B&B) specimen usually is more successful than one transplanted with bare roots; with certain groups this is the only successful method of transplanting. This practice assures that an adequate culture of the proper fungus is established along with the tree or shrub. Furthermore, it has recently been demonstrated that many "bare root" plantings as well as nursery plots are more successful if a culture of the proper fungus is supplied; of course the soil also must be of a requisite type to permit the further growth of the fungus. In many instances when we incorporate humus and put mulches on our gardens we are only secondarily assisting the plants we are growing; the basic function of these items primarily is to set up soil conditions wherein the associated fungi can grow best. Healthy mycorrhizal fungi (and other beneficial soil microörganisms) assure us of good growth in our forest and garden plants.

and those prone to disease, to produce the seed for the next generation. It never seemed to occur that one cannot continue breeding from the scrubs and maintain a healthy and sound forest population with desirable growth habits and resistance to disease. Foresters have only recently realized that an understanding of genetics and good breeding methods is quite as necessary in the production of quality pines as the farmer long ago learned it was in pigs.

The breeding of forest trees is only in its infancy. It is not a spectacular process, for the breeding of trees, where it may be decades between generations, requires the setting up of long-term projects. One of the things in which some of us are particularly interested is in devising ways and means whereby this period of waiting may be reduced or circumvented. It is not an easy task, but there is much more than a mere glimmer of light ahead. I am fully convinced that the next few years will see a great revolution in our silvicultural practices and that soon we shall be reforesting only with the best seed stocks, often of hybrid origin. Hybrid trees are no longer a fantasy of the breeder, but already an actuality.

But, even as in hybrid corn, these trees will make additional demands on the fertility of the soil. It will no longer be considered economic to plant the young trees and then forget about them until time to harvest. We shall have to make certain that the soil is maintained in ample fertility so that these trees can produce their cellulose with maximum efficiency. And let us not suppose that a fast-grown tree will produce weak or inferior wood. It has been amply demonstrated that weak wood is not the result of rapid growth, but of intermittently rapid and slow growth, producing zones of different qualities within the same piece of timber. For all practical purposes, rapidly grown lumber, if the tree is maintained in equal vigor throughout its development, is quite as strong as that which is produced by a slow-grown tree of the same sort.

But it is not enough that we carry on a greatly expanded program of reforestation. Various pests and diseases beset our forests. Within the lifetimes of many of us we have witnessed the loss of our once-important stands of chestnut. Like a great fire, once it got loose, the introduced chestnut blight swept through our forests, so that today we have almost none of these noble trees left. Some would encourage us by hoping that perhaps the few remaining trees may be immune to the disease and so be a source of resistant stock for reforestation. There is some hope of this, but only little. So far, every tree tested has succumbed when inoculated with the blight organism; apparently it was only chance that such trees were not infected during the first cycle of the epidemic.

We all know of the damage caused by the Dutch elm disease and how, in certain sections, it is decimating our elm trees. Recently we have had cause for further alarm, for phloem necrosis, another disease of this tree, has broken loose and is taking a terrific toll of the elms of the midwest. The ultimate fate of the elms can only be conjectured.

About 1920 an infirmity was noted among the beeches of Nova Scotia which later became known as beech bark disease. It was found to be the result of a dual attack, first by a species of woolly scale insect, which weakens the trees but does not kill them, and subsequently by a fungus, which enters through the lesions caused by the insect. This fungus then burrows its way under the bark of the trees, causing girdling cankers which soon kill its host. In some areas, only a decade after onset of the disease, up to 50 per cent of the trees in a stand have been killed.

For some time it was thought that the beech bark disease might be naturally localized in Nova Scotia, and so nothing was done to stamp it out. More recently it was found to have passed the narrow neck of land between Nova Scotia and New Brunswick, where the beeches are considerably more abundant. It was then soon discovered in Maine, and in parts of northern New England it now is completely out of hand, the only practical measures being the felling of all beeches and the conversion of the site to coniferous growths so as to assure a continuous supply of wood products from the area. The most recent information at hand is that the beech bark disease now is present in upstate New York, and still spreading. There is grave possibility that we are now witnessing the destruction of this magnificent forest tree.

And to these we now must add the oaks. In the 1930's it was realized that something was decidedly wrong with the oaks of the upper Mississippi Valley, for many were dying from some unknown cause. Study of the situation revealed that they were being destroyed by a fungus which had never before been known to forest pathologists. Until recently the majority of workers were little concerned with the situation; as usual they somehow imagined that the disease would be naturally confined to a particular geographical area. It was not until early in 1950 that any general attempt was made to assess the situation. It was soon discovered that the Chalara oak wilt, as it is now known, was on the move and spreading rapidly. It had moved out of Wisconsin, Minnesota, and Iowa, and already had gone deep into the Ozarks, and was discovered there in our last remaining stand of virgin white oak timber of any extent; it had crossed Missouri and cut deeply into Arkansas. It was found to be more prevalent among the red and black oaks of Illinois than we had dared suppose, and was later identified as being commonly present in the oaks of northwestern Indiana.

At this juncture, some still maintained that there was no real cause for alarm. Yet before the summer was over, a new infection was discovered near Cleveland, Ohio; and before the autumn closed in and stopped field study, a new locus of infestation was discovered in the geographical center of Pennsylvania, at a point where further infection could sweep throughout the entire length of the great hardwood forests of the Appalachians, mainly composed of oaks, without interference by any known natural barrier.

This new forest epidemic has broken upon us with such suddenness that as yet we do not know how the spores of the causal organism are scattered over long distances to produce further foci of local infection. One dare not predict with confidence what the future of our American oak forests will be. Perhaps something will yet intervene, perhaps the infestation will not reach such epidemic proportions eastward as it has in parts of the upper Mississippi Valley. Those of us who know what it has done to certain oak stands there shudder to think of the possibilities if this scourge becomes widely established among our eastern oaks. It is enough to note that no oak species yet tested, either native or introduced, has been found to be immune.

The chestnuts, the elms, the beeches, and now the oaks. It is a sorry tale I have unfolded to you, and I get no satisfaction in its telling. The time to stop these forest epidemics is when they start, not after they are widely established and out of hand, so that it seems almost physically and economically impossible to bring them under control. I would say that everything possible should be done to check the present epidemics among the elms, the beeches, and oaks, and if possible to confine them until they "burn themselves out." But if this is impossible, if we are witnessing the passing of these important elements of our present deciduous forests, there is no point in wasting time moaning over their loss.

To keep our civilization healthy and intact it is vital that we maintain a constant supply of lumber and wood cellulose. Therefore, if it be necessary, let us start immediately to do the best we can with what we have at hand. Several of us have our eyes on certain exceptionally vigorous and sturdy stands of natural hybrids between species of ash in one of our southern states. One has the feeling that what an errant breeze or casual bee can do the breeder of forest trees can do better. Controlled hybridity among the pines is now an established fact, with promises of the production of sound merchantable timber in one-third to one-half the time required by the natural species. The maples are still with us. The ashes, the buckeyes, catalpas, walnuts, butternuts, and hickories are available. With them are the honey locust and black locust, the sassafras, the mulberry, osage orange, sweetgum, sycamore, tulip tree, magnolia, and cucumber tree. The tupelos, holly, a series of basswoods, the widespread cottonwoods and poplars, the birches, willows, cherries, and others also are at our command. And the great group of cone-bearers, the pines, spruces, firs, hemlocks, larches, cedars, and the rest, are still with us. With these are a host of foreign species which have never been given a completely fair chance to display their qualities as forest crops in this country. If it come to pass, let us regret the loss of others of the best of our native hardwood species, as we must the great stands of chestnut. But let us not permit sentimentalizing about them to interfere with our judgment or throw us into panic.

Forests are not something which have remained static,

either through space or time. The record revealed in the rocks indicates that since early Paleozoic time, species and whole groups of forest trees have come into being, enlarged their ranges, and then passed out of existence. If, by some curious chance, we happen to be living in an age when certain kinds of trees are on the wane, let us make the best of it. Let us valiantly face the fact that sentiment will not save our forests, that positive, constructive action is the only way to their salvation, and that if wood and wood cellulose are necessary things—and they appear to be—then let us produce them by whatever means possible and from such kinds as are available or can be genetically synthesized by the tree breeder.

Let us also be certain to profit by our past mistakes and dalliance and be prepared immediately to stamp out any pest or disease which threatens to curtail the productivity of our forests. To perpetuate any sort of forests, no matter what kinds of trees they may contain, it will be necessary that we treat them as cultivated crops and expend the same relative amount of time and energy in their production that we do on our food-producing grains and vegetables. If by ill chance we cannot save the remainder of our elms, beeches, and oaks, let us be done with weeping. Let us roll up our sleeves to the mighty task before us and get on with the business at hand, for it takes a long time to grow a tree.

THE CONTROL OF PLANT PESTS. Only a few decades ago the food producer had scarcely more than a half-dozen items at his command to combat the various pests and diseases that beset his plants. With these he usually achieved reasonable control. Today the number of such substances is legion. We are prone to boast that this signalizes our recent great advances in plant technology. Before we become too expansive, however, before we burst our bubble of pride, let us critically examine the situation.

Some years ago it was noted that the usual arsenical sprays no longer were controlling codling moth in the orchards of Virginia. Upon further investigation it was discovered that this insect in some manner had dipped into its pool of variant genes and come up with a combination which gave it resistance to this violent poison in concentrations which dared not be sprayed on apples; in brief, the codling moth had evolved a race immune to relatively large amounts of arsenic. It therefore was necessary to discover a new kind of poison to which this orchard pest was not resistant and which might control it.

More recently, when DDT was discovered, we shouted that here at last we had the real panacea (we humans always are hunting panaceas!) for we thought it was to give us full control over a host of garden, orchard, and field pests far too long to list. We now realize that after only a few years' use of this substance, a goodly percentage of these insects (among them the common house fly and the irritating mosquitoes of our summer evenings, as well as the deadly carriers of malaria) are able to sort out genic combinations which produce individuals resistant to its action, just as the codling moth became resistant to arsenicals. The chemist, supplying us with this ever increasing list of compounds, is hard pressed to keep ahead of the situation. We have not understood the nature of the pool of genes inherently at the command of species of living organisms, or the rapidity with which they could shuffle the gene combinations about and evolve new forms. As with the insects, we are discovering that plant diseases caused by fungi in some instances are evolving new forms about as fast as the plant breeder is producing resistant types.

One cannot predict what the end will be. There probably isn't an end, but rather the prospect of a constant war between insects, fungi, and man for the food man raises for himself. Man is an uncommonly stubborn organism, and so keeps to his accustomed routine ways. There are, however, some among us who are beginning to wonder whether new and more deadly dusts and sprays are the ultimate answer, for they now are in such concentrated form that they often cause serious problems in processing food for human consumption. It may well be that we should develop the newer approaches increasingly along the lines of biological control through greater understanding of natural biotic balances. In my estimation, this is the great hope for the future. In the meantime, let us spray.

SUBMARGINAL LAND AND IRRIGATION. At the present time we already are making considerable use of submarginal and arid areas. It is my opinion that we should develop a special type of agriculture in submarginal areas, especially since there is an ever-present danger that they become too dry for any sort of food production and so become dust bowls and ultimately desert wastes. Only under exceptional conditions should they be used for anything except limited grazing. Already we are overgrazing much of our steppe-like grassland areas, so that introduced or even native pest weeds are rapidly on the increase and causing great concern among the animal producers. The fact that various of these weeds also are poisonous to livestock makes an understanding of this situation vital. It is a problem in basic ecology.[4]

[4] In a recent paper on the "Increase of Poison-bitterweed *(Hymenoxis odorata)* on Texas Range Lands" *(Field & Laboratory,* Vol. 19, 29-44, 1951), V. L. Cory notes that overgrazing has led to a serious increase in this stock poisoning plant. In his counts some areas had as many as 2.5 million or even 7.5 million seedlings per acre in the spring; through competition, these might be reduced to only 750,000 or 500,000 mature flowering plants per acre later in the year. However, a single vigorous plant was estimated to be capable of maturing as many as 200,000 seed for the next year's crop of seedlings.

Irrigation of arid lands is often brought forward as a ready panacea. I have never been able to convince myself that irrigation is more than a palliative. The proper use of irrigated soils is one of the more difficult of our agricultural problems and requires a profound knowledge of the physical chemistry of these soils. Considerable publicity is given to the opening of new irrigation projects; for some peculiar reason, we never give the same amount of publicity to their abandonment because of faulty knowledge of the basic physical chemistry of the soils involved. I have not said that we should abandon irrigation as a method of bolstering our food supply. But I do say that it is futile to set up an irrigation project without first determining what the consequences of the irrigation will ultimately be, and also whether there is an ample supply of water. It is remarkable that these items are all too often

---

Noting the increase of other weed types, Cory remarks: "A high proportion of Texas range lands is so infested, and in some parts of the State all of the ranges have one or more species of bromeweed [*Xanthocephalum*]. I have seen entire ranges where the cover of bromeweed hid the growth of all other herbaceous vegetation. At a time not remote these ranges were all in grass." In another passage he notes that "Various other weeds, poisonous or unpalatable forbs, have increased (mostly within the past 30 years) over the ranching country of Texas. Overstocking of the range that results in depletion of the grasses and extermination of the palatable forbs [herbaceous range land plants which are not grasses], leads to its occupation by weedy plants." Revealing as this is, it should be noted that, on the whole, Texas range lands are in better condition than certain others of our western grazing area. The rapid spread and establishment of the saltweed, *Halogeton glomeratus,* a vicious killer, throughout much of the sheep-grazing area of Nevada, and its further spread into parts of California, Idaho, Montana, and Utah, can be laid almost directly to overgrazing. Five million sheep are now being precariously pastured on infested land, with the weed rapidly spreading into previously uninfested areas. A thing which we have not understood is that the effects of overgrazing not only are progressively cumulative, but may not be reversible. Each year for about a half-century we have been sowing a tiny breeze of mismanagement on our natural grazing lands; we now are on the verge of reaping the whirlwind.

overlooked by those engineers who advocate such projects.[5]

TROPICAL AREAS. By tropical areas, I refer particularly to those which normally would be occupied by tropical rain-forest. At the present time there is among us a host of irresponsible persons who advocate the immediate opening up of "the vast unused areas of the tropics" to large-scale agricultural operations. Agriculture in the tropics has little in common with agriculture in temperate regions. Because of constant leaching of the mineral elements under tropical conditions, and a wholly different set of soil micro-organisms (so essential to agriculture in temperate regions), the problem of plant nutrition is a major limiting factor

5 Until fairly recently, each year the Nile overflowed its banks, and this flood water first dissolved the excess and harmful salts which had accumulated in the cropland soil the previous year (as is characteristic of irrigation in arid areas), and then deposited a fresh layer of rich silt. It was this rhythmic pulsing of the river which made systematic agriculture possible in Egypt for so long a time. The system of dams now on the Nile has considerably changed this millennia-old pattern, and the newly projected ones will alter it further, for the dams will hold back much of the flood-water, and the fertile silt certainly will be deposited back of the upper dams and so, ultimately, also render them useless as reservoirs. Another portion of the present project is at least the partial drainage of the Sudd. There is no doubt but that this will throw more water directly into the channel of the White Nile, but others than the engineers concerned are dubious about the ultimate effect on surrounding regions. The drainage of this vast swampy area may well bring about unsuspected and perhaps unwanted secondary effects through changes in soil-water conditions and precipitation cycles. The headwaters of the White Nile are primarily in tension-zone savanna country with an annual precipitation of from twenty to perhaps fifty inches, not a large amount for equatorial regions. Is man, in his attempt to gain another 1.5 million acres of irrigated land in Egypt about to make increasingly arid a considerably larger area in Central Africa? Perhaps only time will tell. Meanwhile, in another part of the world, improperly controlled irrigation is bringing about considerable hardship. Soil and water-table troubles now besetting the great irrigated areas of the Punjab, the bread-basket of Pakistan, are leading to the abandonment of certain parts of the system at an alarming rate; estimates vary, but some observers think that, if not checked in some manner, the loss of crop producing area there may soon reach as much as 100,000 acres per year. Such a loss bodes nothing but ill for this new member of the society of nations.

to the further extension of tropical agriculture at the present time. We must face the fact that, except for specialized, high-profit crops, it is today economically impracticable to supply adequate fertilizers throughout most of the potential tropical crop-producing areas of the world.

Those "do-gooders" who say glibly that they are about to assist the "backward peoples" of the tropics by introducing our own advanced agricultural methods and machinery proclaim nothing but their abysmal ignorance of tropical soil conditions. The tropical "bush farmer" is not being backward or shiftless when he clears a patch, burns the trees, cultivates it for a few years, and then abandons the area. By cutting and burning the forest he is supplying mineral nutrients in ready form, among them potash, so desperately needed in that type of soil. Within a few years the available minerals are used up and he is forced to abandon the land. It then takes another twenty to fifty years, or in some areas as much as a century, for the natural vegetation to come in and deep-rooted trees to "mine" enough nutrients from the earth layers far below the leached zone, and store them in their trunks and branches, so that the trees can again be felled and burned to produce enough mineral nutrients for another brief period of agricultural crops. That is the necessary cyclic pattern of agriculture in much of the rain-tropics at the present time. An extension of the recently developed and more economic method of spraying the mineral nutrients directly on the leaves of the plant, rather than putting them on the soil where they are mostly washed away, may go far in opening up new tropical areas to large-scale agricultural development. But as yet we have not satisfactorily solved the problem of keeping these substances on the leaves during extended periods of rain, and the one thing that characterizes much of the tropics is a heavy rainfall.

Underneath all of this is the ever-present problem of lateritic soil, the bane of the majority of tropical regions. Here is an extremely difficult and involved problem and one wherein the physical chemist may yet make his greatest contribution, for it is one primarily of controlling the colloid balances within these soils. It is not only that these soils are highly leached of their nutrient elements, but any disturbance of them, as happens in agricultural operations, produces a physical change which often makes them unfit for the raising of crops. The nearest I can come to describing briefly what sometimes takes place is that, after clearing and tilling the land, within a year or so a physical change often sets in during which the soil becomes something of the general consistency of wet concrete which is just on the verge of "setting." In fact, under unfavorable conditions, lateritic soils sometimes do "set" so that plowing and planting are impossible. Already we have done much to understand the factors involved, but so far the cost of control too often exceeds the potential value of the crop to be produced. A simple and relatively inexpensive method of handling certain of these more refractory lateritic soils would do more to develop the tropics than any program of betterment yet proposed.

THE LAST RESORT. It is obvious, if we are to supply the world's peoples with adequate nourishment and basic plant materials for the development of a high level of culture, that improved methods of plant breeding will go far in alleviating the conditions caused by our present world-wide cropland shortage, a real and acute situation. But to achieve this potential increase in production, it will be necessary that we also supply a cheap and effective series of mineral nutrients and make considerable advances in land use and erosion control. We must also make better use of our food supplies. There will have to be great advances in our technological applications, and we shall have to cut out all waste of plant materials. At the same time we must venture into a program of forest betterment that staggers the imagination. Irrigation will help, even if it makes up only a minor part of our needs. Furthermore, there is every reason to expect that the near future will see adequate basic research done on tropical soils whereby it will be possible to carry on the amount of agriculture which these regions ought to be capable of supporting.

But human populations, if properly fed and otherwise cared for, seem continually to be on the increase. It is not difficult to envision a time when, with all these measures in complete operation, the population will overtake the available cropland—and we shall be in as bad condition as before. Thus unless control of population increment takes place it ultimately will be necessary that we tap other reserves. It is not boasting, but a simple statement of fact, when I say that the plant scientist is already thinking ahead and getting set for this last eventuality.

Foods are resolved into basic carbohydrates, fats, and proteins, together with the accessory vitamins and minerals. Given a supply of simple carbohydrates and a few minerals, the yeast plant can synthesize prodigious quantities of protein and other valuable substances. This morning many of you took your daily vitamin pill, important ingredients of which were likely derived from these minute fungus plants. A variety of the foods you have eaten today may well have been fortified with its valuable proteins; the excellent and highly nutritious brown gravy on your potatoes, more likely than not, was made almost entirely from refined yeast proteins. If you had a cocktail before dinner, perhaps only as a symbol of the amenities of civilization, the basis of this potable was only a by-product of the metabolic activities of this same plant. Once we come more fully to rely on the yeast plant as a protein producer, the price of raw alcohol will drop to a point where its excellent qualities as an industrial fuel can be more fully utilized.

But from whence are the basic carbohydrates and fats to come? Just recently we have taken certain of the microscopic green algal plants out of our laboratory test tubes and put them to pasture. All that is needed for them is a shallow tank of water exposed to the rays of the sun, the series of minerals required by all plants, and that air, perhaps fortified with a little additional carbon dioxide, be bubbled through the mixture. The result is a greater production of basic carbohydrates and fats than that produced by an equivalent acreage of the best hybrid corn.

At the present time these vat-cultured carbohydrates, fats, and proteins may not be easily translatable into our notions of bread and butter, ham and eggs, or steak and french fried potatoes. But they are basic food substances, and because of them I do not as yet see ahead on any distant horizon the darkening shadow of a Malthusian twilight and the stagnation and end of the race.

SCIENCE AND CIVILIZATION

As we now come to the end of this year's series of Benjamin Franklin Lectures, let us briefly turn the light of history on past civilizations and ascertain whether they can teach us any lesson. Let us put away sentimentality and see what was really at the basis of their culture. I do not intend to classify them, to delineate their wars, or describe their political backgrounds as historians are wont to do. It will be enough to note that, without exception, every one of the great civilizations of the past was developed within a so-called upper class supported either by a degraded peasant-serf or slave population to which the benefits of the culture did not penetrate and from which it was shut off. This pattern is so completely repetitive that one almost is forced to the conclusion that such a situation is necessary for a civilization of any stature to come into being.

But ultimately there came into the world a strange and new philosophy, one which proclaimed that all men were equally free to partake of the advantages of culture. In the long experience of history it would seem almost that such a civilization could not succeed. But before this new idea in the stream of civilization was to be successful, it was first required that man gain partial release from the never-ceasing labor of tilling the land and from slow hand production in industry. These things, only agricultural technology and machines could do. And it is this type of civilization we have produced.

There are among us certain so-called philosophers who go about proclaiming that our civilization "is about to go on the rocks" and that "man soon will be swallowed up in his technology and become a slave to his machines." What such addle-pated Jeremiahs really advocate is that our civilization return to one wherein the majority of men are serfs or slaves. If the benefits of civilization are to be equally open to all, then all must be equally free to enjoy them. Let us then understand the nature of this new idea in civilization, what it means, and how it is to be fully accomplished.

Civilization is a number of things, not the least of which are things of the spirit. To be truly civilized, man must have time for contemplation and pursuit of the arts. The economy of nature is such that, to partake adequately of the things of the spirit, man must have partial release from his necessary daily toil. And this can come only through advances in agro-industrial technology and the production of machines. For the first time in his history man fully understands the dignity of man and so gropes toward a new kind of civilization. For the first time in his history man has had the facilities at his command to produce a civilization in which machines are slaves, and all men may be free.

I have outlined some of the dangers ahead of this new kind of civilization. It is needful to understand these dangers and ascertain the best methods of circumventing them, even of turning them to our own advantage. To make it possible for man's mind to venture into the necessarily new channels of thought, man must remain free. And his continued freedom depends on constant advances in the physical and chemical sciences, in medicine and genetics, as well as in the plant sciences. This problem of keeping man free to develop this new kind of civilization is the business of science in the world today.